Y0-BQC-625

HISTORY, PHILOSOPHY AND SOCIOLOGY OF SCIENCE

Classics, Staples and Precursors

HISTORY, PHILOSOPHY AND SOCIOLOGY OF SCIENCE

Classics, Staples and Precursors

Selected By

YEHUDA ELKANA
ROBERT K. MERTON
ARNOLD THACKRAY
HARRIET ZUCKERMAN

The Growth of Scientific Physiology

Physiological Method and the Mechanist-Vitalist Controversy, Illustrated by the Problems of Respiration and Animal Heat

G. J. GOODFIELD

ARNO PRESS
A New York Times Company
New York – 1975

Reprint Edition 1975 by Arno Press Inc.

HISTORY, PHILOSOPHY AND SOCIOLOGY OF SCIENCE:
Classics, Staples and Precursors
ISBN for complete set: 0-405-06575-2
See last pages of this volume for titles.

Manufactured in the United States of America

Library of Congress Cataloging in Publication Data

Goodfield, G June.
The growth of scientific physiology.

(History, philosophy, and sociology of science)
Reprint of the ed. published by Hutchinson, London, in series: History of scientific ideas.
Bibliography: p.
1. Physiology--History. 2. Animal heat. 3. Respiration. 4. Mechanism (Philosophy) 5. Vitalism. I. Title. II. Series.
QP21.G63 1975 591.1 74-26265
ISBN 0-405-06593-0

The Growth of Scientific Physiology

The Growth of Scientific Physiology

Physiological Method and the Mechanist-Vitalist Controversy, Illustrated by the Problems of Respiration and Animal Heat

G. J. GOODFIELD

HUTCHINSON OF LONDON

HUTCHINSON & CO. (*Publishers*) LTD
178–202 Great Portland Street, London, W.1

London Melbourne Sydney
Auckland Bombay Toronto
Johannesburg New York

First published 1960

This book has been set in Baskerville type face. It has been printed in Great Britain on Antique Wove paper by Taylor Garnett Evans & Co. Ltd., Watford, Herts, and bound by them

For
DAVID SOMERVELL

Acknowledgements

For permission to reproduce copyright material, the author is indebted to the following: Blackwell Scientific Publications Ltd. for extracts from William Harvey's *De Motu Cordis* and *De Circulatione* translated by K. J. Franklin; Harvard University Press for extracts from *A Source Book in Animal Biology* edited by Thomas S. Hall, and *A Source Book in Chemistry* translated by H. M. Leicester and H. S. Klickstein; Prentice-Hall Inc. for extracts from *Great Experiments in Biology* translated by M. L. Gabriel and Seymour Fogel; Professor Warmington and the Loeb Classical Library for extracts from Aristotle's *De Natura Deorum* translated by H. Rackham and, *On the Soul, Parva Naturalia, On Breath* translated by W. S. Hett.

Contents

Preface

THE status of biology as a science has always been a subject for discussion, and the long-standing debate between 'mechanists' and 'vitalists' in physiology is only one aspect of the more general issue. The aim of this book is to go back to the time when physiology as an autonomous science simply did not exist; and to show how, by the efforts of doctors and chemists as well as working biologists, the special character of physiological problems came slowly to be recognized. Both factual discoveries and arguments about method and interpretation played their parts in this process; and the resulting science preserves something of the attitudes and interests both of the clinicians and naturalists who contributed so much to an understanding of its problems, and of the workers in physics and chemistry who provided the tools for their solution. The problem of animal heat, which is my chief illustration, served as a topic round which raged many of the bitterest disputes concerning physiological method: if one follows out the steps by which this particular problem was solved, one can see how the proper role of physiology in our understanding of life came gradually to be appreciated.

I have many people to thank for helping me with this book. Much of the work was done during my tenure of a Fellowship in the History and Philosophy of Science at the University of Leeds, for which my gratitude is due to the Leverhulme Trustees. The Librarians of the Brotherton Library (University of Leeds), the Royal Society of Medicine, and the Wellcome Historical Medical Library gave me valuable help and facilities; but above all I owe a great debt to Miss Eileen Read, Librarian-in-charge at the School of Medicine, University of Leeds, for giving me encouragement and help at the cost of much time, labour and inconvenience to herself. Miss Dorothy Raper and Mrs. Sara Ligtelyn have helped in the preparation of the manuscript. To Mr. Terence Cawthorne, F.R.C.S.,

Honorary Secretary of the Royal Society of Medicine, I owe a special and personal debt, not only for his constant encouragement and help, which reflected the interest that his profession has always taken in the development of its own methods and ideas, but also for what I myself have gained from his professional skill. Finally, I have had useful discussions with Stephen Toulmin, during which some of the main philosophical distinctions used in this book were hammered out. I am grateful to him for his rigorous criticisms and for his constructive advice, even if I have not always acted upon it.

Leeds and London 1956–9 G. J. G.

POSTSCRIPT

Since this book went to press I have discussed with Mr. Everett Mendelssohn of Harvard University the work he has been doing on theories of animal heat. Since many of our conclusions appear to be similar, it should be stated here that they were arrived at independently and without any consultation. (Our agreement is perhaps a mark of truth, rather than of telepathy!) I understand that Mr. Mendelssohn is preparing a series of papers on the subject, which will be published in the near future. I have not had the chance to read these; nor has he read my manuscript.

Physiological Method and the Problem of Animal Heat

WHAT is life? What is it that makes living things unique? The story of man's attempts to answer these questions provides one central strand in the history of biology. Each new step forward in our understanding of particular bodily functions and structures has called forth a new debate about 'the Nature of Life'; and the scope which we are entitled to allot to biological progress has been re-argued at each stage. But though the terrain of debate has shifted; though the methods of inquiry have changed; and though the working problems of biology have continually moved from one field to another, at each epoch one finds a similar division of opinion, between 'mechanists' on the one hand and 'vitalists' on the other. The first group are confident that the full explanation of biological phenomena can be found by a sufficiently intelligent extension of known scientific principles and laws; and in their eyes biologists in the opposite camp have been 'timid and hesitant', and have tried to interpret biological activities as a 'series of miracles'. (a) The vitalists for their part, have at each stage accused the mechanists of presumption, and worse, of making nonsense of the manifest differences which mark off the behaviour of living things from that of the inert.

The very inconclusiveness of this debate entitles one to ask whether the real issues have been clearly appreciated. Is there perhaps a pervasive cross-purpose which in each generation stimulates a recurrence of the same battle?

One way in which it may be possible to throw fresh light on this vexed dispute is by the study of the development of men's

(a) 6, p. xiv. (All references are to the bibliography on p. 165, e.g. the reference here is to item number 6. On looking up p. xiv in item 6, the quotation will be found.)

ideas about one particular function characteristic of animals, which perplexed biologists for many years; and to see how these changes in ideas are reflected in the debate about 'the Nature of Life'. One function is especially worth picking on for this purpose; and the progressive unravelling of its mechanism forms a principal theme of this book. For when one looks for the obvious signs of the unique character of organisms, one feature will be seen to have caught men's attention from earliest times – the way in which animals maintain their warmth and their breath, despite the fluctuations of their environment. The motions of the life-blood; the breath of life; the maintenance of animal heat – an understanding of these three closely-connected activities has always been seen to lie at the heart of physiology. And in the warmth of the body, kept up – it seemed – by the flow of the life-preserving blood, and perpetually recharged by the *pneuma* in the lungs, lay the key to the problem. Where a corpse, like any inert body, would quickly take up the temperature of the surrounding air, the living body showed its autonomy – and perhaps its very life – by preserving in the face of surrounding fluctuations a steady, and usually a higher, temperature.

The aim of this essay is to trace out the story – reconstructed from the writings of the men concerned – of their ideas about this phenomenon of 'animal heat'. The problem occupied the centre of the stage at a crucial period in the growth of physiology. The story as it unfolds can help us to a better understanding of the nature of the age-old and shifting debate between mechanists and vitalists. And out of the argument between chemists, doctors and professional physiologists, which led by painful stages to a solution of the problem, physiology was eventually to emerge as a science on a par with physics and chemistry.

The word physiology comes from the Greek word *physis* which can be somewhat loosely translated as 'nature', although the precise meaning of this word altered a great deal in the course of the development of Greek thought. Our own word 'nature' is almost as ambiguous. Thus the study of physiology becomes the study of the 'nature of things'; and in the twentieth century we would restrict the term specifically to the study of the nature of living organisms.

This definition presupposes two things: first, that we can recognize a living organism when we see one, and second, that we have accepted ways and means of studying its nature. Here, from the very start, is a difficulty. How are we to know *what is* the nature of a living organism unless we study it; but how can we begin to study it – i.e. decide the sorts of questions to ask about it – unless we know what sort of phenomena we are dealing with? At the outset we have to make an assumption: that the phenomena in question *can* be explained in the kind of terms we are going to use. The early history of physiology is, accordingly, the history of the various ways in which men have set about trying to explain the behaviour and properties of living material; and the most heated conflicts in physiology have chiefly been conflicts about the legitimacy of the resulting explanations – controversies over interpretation rather than facts.

I said that our definition presumed that we could recognize a living organism when we saw one. At a level of sufficient complexity this is true enough, and throughout the history of biology there has been a strand of observational study by means of which the unique activities of organisms have become progressively better known. This aspect of physiology reached a climax with the work of Haller and Hunter; after this one could, in general, mark off a living organism by reference to certain characteristic vital activities, such as growth and reproduction. Yet we cannot simply *define* a living organism as one which is performing these activities. As recent biochemical research shows, one is often unable to say by this standard whether something is living or not; at times it will manifest some, though not necessarily all, of these activities; and at other times it will not show any activity at all. Nor can one give a full explanation of the processes of life in these descriptive terms alone, even though today we may be able to distinguish different types of vital activities at different levels of molecular or cellular complexity.

Moreover, it is unwise to attempt to explain the distinction between living and non-living things in terms of the materials of which they are composed. There may be a difference in organization and complexity at the molecular level, or a difference in dynamic activity, but there is no difference in the basic chemical elements which form the building materials. A biochemist may say that proteins (for all that they can be synthesized in the laboratory) are characteristic in nature of those things that show living activities; but he would never go so far as to say that these organisms were made up of a different kind of carbon, hydrogen and nitrogen from that of inert things. If he did, he would undoubtedly be condemned as something far worse than a vitalist! However, it is not enough just to recognize the fundamental identity of material in living and non-living things – the Greeks, too, rejected the distinction, though from the other side. Many of them believed that those 'unique living activities', such as embryological development, were common to both inanimate and animate things. Chemical substances 'grew' in the same way as biological organisms; an explanation of chemical change was to be sought in biological terms – and this attitude, which can be traced back to Aristotle, persisted right up to the time of Van Helmont and Harvey. In the course of time, therefore, the word vitalism has come to cover a whole range of different ideas. Paradoxically, a modern biochemist would be labelled a vitalist for insisting on a material distinction between living and inert things, which Aristotle did *not* recognize; while Aristotle is in turn also called a vitalist for other reasons. Singer says:

> 'He was before all things a "vitalist". For him the distinction between living and non-living substance is to be sought . . . in the presence or absence of something that he calls *psyche,* which we translate *soul.*' (a)

What then is a vitalist – in the modern sense of the word? He is (it seems) a man who not only *describes* the differences between inanimate and animate material, but seeks to explain

(a) 82, p. 25.

those differences in terms which either exclude inanimate material or else include agencies over and above those recognized in *modern* physics and chemistry. Of course, speaking on a descriptive level, living things *are* different from non-living ones. Though the status of a virus, as we all acknowledge, may be a source of difficulty, no one can fail to see the difference between a rabbit and an ink-pot. But to say wherein this difference lies is another matter. Aristotle's explanation of it is in terms of a non-material agency – the psyche. Later one finds men using the terms 'animating principle' and 'vital principle', and also the 'Archeus' of Paracelsus. Most sophisticated scientists in the twentieth century feel that any scientific explanation which does not rest basically on physico-chemical ideas must be rejected. These ideas must be our starting point in physiology as well, even though later on we may have to bring in concepts of an essentially biological kind in order to complete the picture. But there is no *a priori* reason why this should be so; the belief that an explanation in these terms is justified is an assumption quite as much as any other, and it is – like physics and chemistry themselves – a comparatively modern development. Though to a certain extent anticipated by Descartes, it was never fully accepted until the late nineteenth century, after the work of Claude Bernard. My chief concern is to ask why this was so.

It is not surprising that early attempts to express the difference between inanimate and living material should have been made in terms quite different from our own. William Harvey and René Descartes inherited a biological tradition which was firmly based on the writings (more or less well understood) of Aristotle and Galen. For both men the study of function and development was fundamental, and structure was to be understood in these terms; to explain the nature of a phenomenon it was essential to place it in an 'embryological' situation, as it were – to ask what *sort* of a thing one was studying and what was the *destination* towards which its development was directed. This was true irrespective of the type of phenomenon in question. Physicists and chemists nowadays ignore such developmental questions, concentrating on the conditions under which natural phenomena occur and the mathematical relations holding between them. But biologists, and more particularly physiologists, inevitably retain more of the Aristotelean attitudes. (One is surely justified, when studying anatomy, in asking questions about 'the function' of an organ.) And up to the time of Harvey, physiology dealt more with these functional aspects than with any others.

Nevertheless, early physiology was certainly an experimental science. Galen performed many ingenious and careful experiments–to determine the functions of the kidney and bladder, for instance. Harvey's greatness lay not only in the remorseless way in which he argued that the blood *must* circulate, but also in the convincing experiments by which he demonstrated that in fact it *does* circulate. And, as Wightman(a) points out, he added something new; he used observed structure to explain function, as well as function to explain structure. But functional explanation alone leaves still unanswered questions about the fundamental types of explanation *appropriate* to living things. To say that the bladder is made of an elastic-like material because its function is to store urine – or to say that the heart has elastic walls because its function is to pump blood round the body – is in one sense to give no more than a description. To state what makes the kidney secrete the urine, or what keeps the heart pumping, is another matter. And as for the question of why non-living things do not behave in the same way from a functional standpoint, one could only say, 'Well, they don't have to'.

Now Harvey was in no position to deal with these further questions. Scientists in the seventeenth century still could not draw any clear distinction between chemical and biological processes, between chemical and biological substances, or between physical and chemical change. The use of the word 'spirits' illustrates the underlying difficulty nicely. It could at one and the same time mean a gas, a volatile fluid, a soul, an animating principle, something which was the cause of nervous energy in the animal, something which was the cause of its

(a) 85, p. 5.

being alive, and something which when mixed with various other substances was the cause of heat in the animal. It was a word that carried both material and immaterial, and descriptive and explanatory implications. Harvey stood out amongst his contemporaries in recognizing the ambiguity of the word 'spirits', and how inappropriate were many of the roles it was called on to fulfil.

'With regard to the third matter, namely, spirits, there are many and opposing views as to which these are, and what is their state in the body, and their consistence, and whether they are separate and distinct from blood and the solid parts, or mixed with these. So it is not surprising that these spirits, with their nature thus left in doubt, serve as a common subterfuge of ignorance. For smatterers not knowing what causes to assign to a happening, promptly say that the spirits are responsible and introduce them as general factota. And, like bad poets, they call this *deus ex machina* on to their stage to explain their plot and catastrophe . . .

'In general, however, the school of physicians agree on three kinds of spirits, namely, those of growth permeating through the veins, those of life through the arteries, and those of the psyche through the nerves. Hence the physicians say, after Galen, that sometimes the parts work in sympathy with the brain because an ability is restrained with essence, that is spirit, at other times irrespective of essence. Further, in addition to these three inflowing kinds of spirits, he seems to assert an equal number of stationary ones. I have, however, never found such in veins, nerves, arteries, or parts of living subjects. Some make the spirits corporeal, others incorporeal, and those who want them corporeal sometimes make the blood, or its thinnest portion, the link with the psyche. Sometimes they conceive of the spirits as contained in the blood (like flame in the aroma of cooking) and sustained by its continuous flow; sometimes of the spirits as distinct from the blood. Those who declare the spirits incorporeal have no ground to stand on, but they also recognize capacities as spirits (such as digestive, chyle-forming,[1] and procreative spirits) and admit as many spirits as they admit faculties or parts.' (a)

Yet Harvey himself had to retain the term throughout his career. In the *De Motu Cordis* (1628) he writes: 'No one denies that blood . . . is imbued with spirits'.(b) And twenty-one years

[1] In these cases the 'spirits' provided the explanation for the phenomenon.
(a) 42, p. 37. (b) 41, p. 12.

later in the *De Circulatione Sanguinis* (1649) he goes on to explain the proper sense in which the term was to be used – though the account he gives itself reveals a basic confusion which was to take another century to clear up.

'What, however, is specially relevant to my theme after all other meanings have been omitted from consideration as being tedious, is that the spirits escaping through the veins or arteries are no more separate from the blood than is a flame from its inflammable vapour. But in their different ways blood and spirit, like a generous wine and its bouquet, mean one and the same thing. For, as wine with all its bouquet gone is no longer wine but a flat vinegary fluid, so also is blood without spirit no longer blood but the equivocal gore. As a stone hand or a hand that is dead is no longer a hand, so blood without the spirit of life is no longer blood, but is to be regarded as spoiled immediately it has been deprived of spirit. Thus the spirit, which is specially present in the arteries, and arterial blood, is either the product of such blood, like wine's bouquet in wine, and the spirit in brandy; or like a small flame kindled in spirit of wine and keeping itself alive on such a diet. In consequence, the blood, though very heavily imbued with spirits, is not turgid with them, nor do they cause it to rise or become blown out so that it wants and needs more space (which, in the experiment already referred to above, you will be able to determine very accurately by measuring the vessels): but, like wine, it is to be understood as prevailing by its superior strength and its vigour in doing and effecting, in the Hippocratic sense.' (a)

Harvey's attack on the indiscriminate use of the term 'spirits' is of course entirely justified. At the same time, with chemistry at so primitive a stage of its development, it was difficult to see any other terms in which one could both describe the activities of living things and explain their material constitution. Harvey, realizing this difficulty, is quite clear on one point: that our explanations must begin with factors that we can *observe*, whether these can be seen directly, or have to be demonstrated experimentally. To admit incorporeal factors will not help us.

'Silly and inexperienced persons wrongly attempt . . . to upset or to establish which things should be confirmed by anatomical dis-

(a) 42, p. 38.

section and credited through actual inspection. Whoever wishes to know what is in question . . . must either see for himself or be credited with belief in the experts, and he will be unable to learn or be taught with greater certainty by any other means . . .

'If nothing could be admitted by sense without the evidence of reason, or on occasion against the dictates of reason, there would now be no problem for discussion. If faith through reason were not extremely sure, and stabilized by reasoning (as geometers are wont to find in their constructions), we should certainly admit no science: for geometry is a reasonable demonstration about sensibles from non-sensibles. According to its example, things abstruse and remote from sense become better known from the more obvious and more note-worthy appearances. Aristotle advises us much better when, in discussing the generation of bees . . . he says, "Faith is to be given to reason if the things which are being demonstrated agree with those which are perceived by sense: when they have become adequately known, then sense should be trusted more than reason." Hence we ought to approve or disapprove or reject everything only after a very finely made examination. But to test and examine if things are rightly or wrongly spoken, ought to lead to sense, and to confirmation and establishment by the judgment of sense where nothing false will remain hidden . . . For no knowledge can come save from pre-existing knowledge, and this is one of the reasons why our knowledge about the heavenly bodies is so uncertain and conjectural.' (a)

Now what I take him to mean here is this: it is legitimate in our search for explanations to proceed by the use of either reason or faith; but the sole criterion of whether or not our faith and reason have led us to the right point is whether or not we can then actually see and demonstrate the correctness of our conclusions. If we cannot, then over this point our faith or reason was at fault. (Harvey incidentally was unable to demonstrate the existence of the blood-capillaries, but inferred their presence from the fact that they were necessary for the circulation of the blood. He did, of course, demonstrate here that the blood could not go backwards up the arteries, and that it must flow continually forwards on through the valves in the veins. At this point he reasoned that unless the whole mass of circulating blood were constantly used up and

(a) 42, p. 54.

as constantly replenished, there MUST be connexions between the arteries and veins – but he never saw them.)

Harvey's emphasis on visual demonstration and experiment as a test for ideas forms one strand in the emergence of modern physiology, and one which was never entirely lost, even during the greatest controversies within the subject. The other chief strand, which was more bitterly argued, comes from Descartes. Descartes committed himself to a mechanistic approach far more deeply than Harvey, without at that stage having the same degree of justification. He flatly rejected every explanation of the behaviour of organic structures other than the sort that one would apply to the study of inanimate material. About this he was exact and emphatic. With the exception of the reasoning soul of Man, which in any case he felt he had to place in a *material* location – the pineal gland – all things are to be understood and studied in the same terms.

'I desire you next to consider that all the functions which I have attributed to this [animal] machine, such as digestion, the beating of the heart and the arteries, nutrition and growth, breathing, waking and sleep, the perception of colours, sounds, smells, tastes, heat, and other such qualities by the external senses, the impression of their ideas in the organ of *sensus communis* and of imagination, the retention or impression of these ideas in memory, the internal motions of appetites and passions; and finally the external movements of all limbs, which follow so suitably as well from the actions of the objects presented to sense as from the passions and impressions which are found in the memory, that they imitate as perfectly as possible those of a real man – I desire you to notice that these functions follow quite naturally in the machine from the arrangement of its organs exactly as those of a clock, or other automaton, from that of its weights and wheels; so we must not conceive or explain them by any other vegetative or sensitive soul, or principle of motion and life, than its blood and its spirits agitated by the heat of the fire which burns continually in its heart, and which is *of no other kind than all the fires which are contained in inanimate bodies*.'[1] (a)

This is mechanistic philosophy in two senses, which have to be kept distinct. It is mechanistic in the sense that it emphasizes the machine-like nature of the organism – an aspect of

[1] My italics. (a) 28, pp. 201–2.

mechanistic physiology which was run to death for a while after Descartes and which ultimately provoked an equally violent movement away from it in the Montpellier school of the eighteenth century. But it is mechanistic in another sense as well – in the sense in which the term 'mechanist' is used throughout this book. Descartes implies and insists that living organisms and non-living things conform to the same laws and therefore animate things must be studied with the same techniques as inanimate ones. There is nothing unique about organisms except their *activities*, and no special factors are therefore necessary to explain their uniqueness.

Two hundred years later Bernard was to say exactly the same thing, though by this time there was far more justification for doing so. Yet in Descartes' time physics and chemistry were not as distinct either from each other or from biology as they were in the time of Bernard, so that perhaps the faith he shows in his mechanistic principle is understandable.

But stating a principle is one thing: sticking to it in practice is another. As soon as men began to create a truly mechanistic physiology, the magnitude of the difficulties involved became apparent. For, after all, the differences between living and non-living things are undeniable: notably the very phenomenon of animal 'heat' which is our main topic. The quotation from Descartes about the 'central fire' provides a convenient starting point for our study. The way in which warm-blooded animals preserve a body-temperature which is both constant and well above that of the environment, in spite of external variations, is in striking contrast to the way in which inanimate things rapidly attain thermal equilibrium with their surroundings. Judged by the standard of inert things, the living body seems at first sight to act in defiance of the laws of physics and chemistry. Newton's Law of Cooling, which states that a body cools at a rate proportional to the excess temperature over the environment, seems not to apply to animals in the way it does to all inanimate things. To suppose that any theory of heat could ever be given in terms applicable equally to living and non-living things required at this time a faith that would remove methodological mountains. As it was, the chemical processes responsible for the phenomena of animal heat alone took about two hundred years to sort out (the same length of time as was needed for the Copernican Revolution) – and until these processes were analysed no complete physiological picture could be built up. Many advances in physics and chemistry had to precede the first physiological analysis – to mention only three discoveries: the nature of gas (Van Helmont); the exact kinds of the gases that were respired (Lavoisier, Priestley); and the absorption of respired gases by the blood (Faust and Mitchell, etc.). But the central issue was one of methodology: – whether we are in any case entitled to study living things experimentally, and if so, what such experiments can tell us. The greatness of Claude Bernard lies as much in the way he answered these methodological questions as in his particular physiological observations. In the comprehensiveness of his physiological hypothesis he can justly be said to have played Newton to the Copernicus of Descartes.

2

Theories of Animal Heat before 1800

THE difference in thermal behaviour between living and non-living material had been seen and remarked on as early as Aristotle. The ancients realized that because of the difference in 'hotness' between the animal and its surroundings, 'fire' must continually be lost and that the living organism was yet able to replenish its heat. Fire was one of the four fundamental earthly elements recognized by the Greeks, and we find some of them using the term 'animal fire' synonymously with 'vital principle'. This living fire was untouched by any physical agency. It was an innate quality or property of the body; something essentially connected with life. Cicero, quoting the doctrine of Cleanthes, wrote:

> 'It is a law of Nature that all things capable of nurture and growth contain within them a supply of heat without which their nurture and growth would not be possible; for everything of a hot fiery nature supplies its own source of motion and activity; but that which is nourished and grows possesses a definite and uniform motion; and as long as this motion remains with us, so long sensation and life remain . . . Every living thing therefore, whether animal or plant, owes its vitality to the heat contained within it. From this it must be inferred that this element of heat possesses in itself a vital force that pervades the whole world.' (a)

So struck were the Greeks by this generation of animal heat that they speculated less about its cause than about the ways in which it could be controlled, so that no harm should come to the animal.

The source of this fire, according to both Aristotle and Galen, was the heart: it was carried along the arteries to the rest of the body together with the 'pneumata' or 'spirits'. One of the functions of the movement was to distribute the heat from the

(a) 21, pp. 23–8.

heart; that of the respiration – a term for a long time used synonymously with breathing – primarily to cool the blood. For instance, Aristotle said, in his essay on respiration:

'As for animals with blood and a heart, all that have a lung admit the air and achieve cooling by breathing in and out . . .

'. . . The reason why those that have lungs admit the air and breathe, and particularly those which have a lung charged with blood, is that the lung is spongy and full of tubes . . .

'. . . All creatures that have this part charged with blood need rapid cooling because there is little margin for variation of their vital fire, and air must penetrate the whole lung because of the quantity of blood and heat which it contains . . . the breath passes easily to the source of heat which lies in the heart . . .

'. . . The source of life fails its possessors when the heat which is associated with it is not moderated by cooling; for then the heat is consumed by itself . . .' (a)

To pass to the modern period: it was early suggested by Servetus that the inspired air might have uses besides that of refrigeration.

'The vital spirit has . . . its source in the left ventricle of the heart . . . It is a fine attenuated spirit, elaborated by the power of heat . . . engendered by the mingling of the inspired air with the more subtle portion of the blood.' (b)

The cooling function of respiration was believed by some to be augmented by the pulsation of the heart and arteries, though Van Helmont, seeing that the action of a bellows increased the flame of a fire, argued that the function of the pulse was to intensify the heat, not reduce it.(c) Harvey also, for other reasons, threw doubt on the cooling function of arterial pulsations, and even – though this is often overlooked – questioned whether the heart is really the source of heat.

'If the arterial pulsations cool and fan the parts of the body as the lungs do the heart itself, how is it commonly stated that the arteries distribute from the heart to the individual parts blood packed with the spirit of life, which foster the heat of the parts, rouse it when

(a) 4, pp. 455, 467, 469, 473.
(b) 81, p. 206, Willis. (c) 43, p. 180.

torpid and, so to speak, restock it when low? And how if you ligate the arteries do the parts not only straight-way become sluggish, cool and turn somewhat pale, but also finally cease to be nourished? According to Galen this is because they have been deprived of the heat which had previously flowed to all parts from the heart, since it is clear that the arteries transmit heat rather than cooling and ventilation from the heart to the parts. Moreover how in diastole can there simultaneously be drawn spirits from the heart to warm the parts and from the outside, means for their cooling? . . . Such views are clearly mutually conflicting and contradictory.' (a)

Twenty years later he still maintained this view.

'It is also to be noted . . . from those . . . who believe also that the spirits and inflowing vital heat arise from the heart (and this by the innate heat of the heart, as if by the instrument of the psyche, or common bond, and prime organ for carrying out all the operation of life); who thus think that the blood, and the movement of spirit, perfecting and also warmth, are borrowing from the heart as from the source . . . If I may speak openly I do not think these things are so.

Nor is the heart, as some think, like a sort of burning coal or brazier or hot kettle, the source of heat and blood, but rather the blood, as being the warmest part in all the body, gives to the heart . . . the heat which it has received.' (b)

He asserts that the heart is like a hot kettle only because the blood it receives is already hot, and not because it is itself the source of heat. Yet he too held the view that the lungs cooled the blood; and this belief can be found right up to the end of the eighteenth century.

'For as the inspired air tempers the excessive heat in the lungs and the centre of the body, and looks after the expulsion of suffocating fumes, so in turn does the hot blood, sent through the arteries to the whole of the body, warm all its parts . . .' (c)

Nowadays we realize that theories which postulate a focus of heat-production must allow for a temperature-gradient outwards from the source, unless the heat is carried in some bound or insensible form. So long as there was no clear distinction between heat and temperature, or between the qualitative

(a) 41, p. 11. (b) 42, pp. 62, 63. (c) 42, p. 20.

and quantitive aspects of heat, this necessity could not be recognized. And since one can actually *feel* the heat of the body, one would not immediately suppose that it might be carried internally in an insensible manner.

By the middle of the eighteenth century we find a number of serious attempts to explain the *cause* of animal heat: this debate culminated in the writings of Lavoisier and Laplace (1783) and Adair Crawford (1779). In the years preceding their work, four different sorts of explanation were put forward. The two we shall consider first were mechanical, attributing the heat of the animal either to friction between the blood and the walls of the blood-vessels, or to friction between the solids in the blood itself. These mechanical views were derived from the ideas of Borelli and Boyle, and they were also held by Hales and Haller. The ideas were based on the observations that exercise increases the heat of an animal and that movement against friction produces heat. Haller observed the movement of the blood in the vessels and the change in shape of the 'globules' as they were forced through the capillaries.

'The red part of the blood seems chiefly of use to generate heat, since its quantity is always in proportion to the heat of the blood. This being confined by the largeness of the globules, within the red and first order of vessels, hinders them from collapsing; and, in receiving the common motion of the heart by the greater density of its parts, it has a greater impetus . . . The globular figure of its parts, together with their density, makes it easily pervade the vessels; and the quantity of iron it contains, as well as of oil, perhaps increases its power of generating heat . . .

'We may also ask, whether the heat of the blood does not also proceed from its motion? . . . Is not the truth of this sufficiently evinced by the blood's being warm in those fish which have a large heart and cold in such as have a small one? the generation of heat being in proportion to the size of their bodies: . . . from the increase of animal heat that ensues from exercise of all kinds, and even from the bare friction of the parts?' (a)

In view of the role we now know that iron salts play in the red corpuscles, Haller's ideas are delightful. As is shown by the

(a) 40, p. 106.

quotation above, he maintained that the chief use of the 'red globular parts' of the blood was to generate heat, and he considered that the iron present was 'matter very fit to enter into the vibrations necessary to the production of animal heat'.

According to Bostock(a) the last attempt to form a mechanical theory of animal heat was made by Douglas in 1747, when he declared 'that animal heat is generated by the friction of the globules in the extreme capillaries'.(b) Bostock is probably right; a study of the important literature seems to show that no mechanical theory was held after 1750, though a century later Joule himself was tempted to revive the idea.(c) Mechanical views were not easy to substantiate, as Leslie points out: it was difficult to give a single convincing example of heat being generated by the mutual action of a solid and a fluid body. Two analogies were usually relied on; that of a cannon ball going through the air, and that of 'quicksilver strongly agitated in a glass phial'; but neither of these was sufficient illustration. In the first case, Leslie argued, it was more probable that the heat of the ball was due to the initial explosion of the gunpowder; and in the second, heat was not generated unless a portion of the mercury was 'transmuted into a powder', so that it might equally well be due to the rubbing of a solid against a solid.(d) Douglas saw that single globules in the blood could only pass through the capillaries one at a time and therefore

'. . . in those vessels the fluids are reduced to such small particles that the attrition may be considered as that of solids acting upon solids.' (e)

Leslie rejected this view on the rather uncertain grounds that

'. . . we can perceive no foundation for accounting one of those red globules a solid body; granting that the division to be the minutest possible, those particles are by no means atoms, but still masses of fluid and of so soft and flexible a texture, too, that they are universally allowed to be capable of elongation.' (f)

(a) 16, p. 438 footnote. (b) 29, p. 27. (c) 49, p. 442.
(d) 58, p. 58. (e) 29, p. 12 (f) 58, p. 62.

But on this point of observation Douglas was right; amoeboid shape and movement belong to white blood corpuscles only, and red corpuscles are solid elastic discs.

But the greatest objection to these mechanical theories arose over the amount of friction they called for. It was realized that heat would be produced only if there was enough friction, and that this in turn depended on the roughness of the moving surfaces, and could be reduced by suitable lubrication. Yet it would be difficult to find surfaces so perfectly lubricated as those found in blood vessels and between muscles, so where did all that friction take place? This objection was equally serious whether one regarded the attrition as being between solid and liquid, or between solids. Both mechanical theories were therefore equally affected, and neither was able to establish itself at all widely.

The third of the four theories current in the seventeenth and eighteenth centuries stated that animal heat was produced by some kind of fermentation or chemical mixture. Views of this general type persisted at least up to 1778. Early proponents of the mixture theories were Van Helmont and the French anatomist Sylvius. Van Helmont believed that the heat arose from the mixing of sulphur and volatile salts of the blood; Sylvius thought that it was formed by the effervescence of humours created when blood and chyle met. The discovery that heat was produced when acids were mixed with alkalis suggested that animal heat arose in the body in a similar way. Fermentation theories were possibly more fruitful than mixture ones since they had direct observations to help them; both ferments and putrefying substances can be seen to generate a considerable quantity of heat.

About the same time as Black was relating his discoveries on carbonic acid to respiration, Benjamin Franklin threw out a conjecture about animal heat. He was troubled by the fact that the mechanical theories were inadequate because the movements of fluids on solids did not produce heat. Knowing that animals must constantly make good the loss of heat to their surroundings, he says:

'I am inclined to think that the fluid fire, as well as the fluid air, is attracted by plants in their growth and becomes consolidated with the other materials of which they are formed and makes a greater part of their substance: and that when they come to be digested, and to suffer in the vessels a kind of fermentation, part of the fire, as well as part of the air, recovers its fluid active state again and diffuses itself in that body digesting and separating it: that the fire so reproduced, by digestion and separation continually leaving the body, its place is supplied by fresh quantities arising from the continual separation. That whatever quickens the motion of the fluids in an animal quickens the separation and reproduces more of the fire; as exercise. That all the fire emitted by wood, and other combustibles, when burning, existed in them before, in a solid state, being only discovered when separating . . . and in short, what escapes and is dissipated in the burning of bodies, besides water and earth, is generally the air and fire that before made parts of the solid. Thus I imagine animal heat to arise by or from a kind of fermentation in the juices of the body, in the same manner as heat arises in the liquors prepared for distillation, wherein there is a separation of the spirituous, from the watery or earthy parts. And it is remarkable, that the liquor in a distiller's vat, when in its highest and best state of fermentation as I have been informed, has the same degree of heat with the human body; that is about 94 or 96.' (a)

Here we also have one of the earliest suggestions that heat could be carried in a bound, and therefore inactive, state; and that burning restored it to 'its fluid active state'.

Franklin's suggestion was taken up by Rigby, and expanded into a theory of animal heat which he published in 1785.(b) This is about the last 'fermentation' theory of animal heat to be printed, and already a new element has been introduced, for Rigby clearly realizes the importance of the digestion. He briefly rejects all 'common, sensible and external' sources of heat as being inadequate: the immediate source must therefore be latent and internal, and this rules out the atmospheric air as a possible source of supply.[1] Using 'the doctrines of

(a) 38, pp. 79 and 125. (b) 80.

[1] Rigby's essay was published well after Priestley's work on fixed air. Black's work on latent heat and temperature and his theory of animal heat, the first edition of Crawford's book on animal heat, and Lavoisier's paper on respiration and combustion. Admittedly Rigby's work was published as an

latent heat', Rigby puts the view that animal heat arises from the decomposition of food which would, in the combined state, contain a considerable quantity of heat in a fixed form. Using the contemporary chemical definition of digestion as 'an intestine motion arising spontaneously among the insensible parts of the body, producing a new disposition and a different combination of those parts', he goes on to affirm that this process involves a sensible production of heat. He uses as confirmation of this fact 'the sense of superior heat which is obviously felt in the region of the stomach'. From the indigestible quantities of food that people were accustomed to eat in those days, his contemporaries would probably agree that this was indeed direct evidence that could not be lightly ignored. This decomposition, he states, will take place only if 'the balance between the several parts' which 'in their fixed and perfect state depend on an exact and proportionate combination of their several ingredients' is upset.

'Thus supposing water and the matter of heat to be two of the principal ingredients which enter into the composition of a grain of barley, if a super-abundance of these is given it, the exact proportion between them and other materials forming the body will be taken away . . . In the same way food when in the stomach is "afflicted" with heat and moisture and undergoes the most perfect decomposition and rearrangement of parts and the matter of heat which probably forms a very considerable part of them should be separated and rendered capable of diffusing itself through every part of the body . . . And if we trace the blood through all its changes and consider that it

---

essay and he may not have felt that space was available for a consideration of other ideas, but he makes no mention of any other theories or his reasons for rejecting them except to say that they do not satisfy him.

'I have not thought it requisite to mention any of the arguments, which I think might easily be adduced against the several theories before alluded to, as the mere disproving them would tend but little to confirm the present; and as, should the present be considered as a rational one, it would require no support from such a circumstance.' (a)

One may be surprised at his attitude, especially in comparison with the extremely careful way in which Leslie examines all the prevailing theories. There can be no question that he knew contemporary thought on this subject; there is a full account and criticism of Crawford's work, by Morgan, in the same monograph in which Rigby's essay appears; in fact immediately before it.

(a) 80, p. vii.

undergoes several new arrangements in its progress throughout the different parts of the body then these new decompositions will also produce heat. The new arrangement of their component parts will retain heat in the concealed state and every such change [decomposition] must be productive of sensible heat. Heat production is not then localized in any one place nor is heat stored in any one place. The formation of fat inevitably means the storage of heat. I cannot help feeling that the prevalence of corpulency among Dutchmen is in some measure owing to the retention of heat; for though they are not a slothful people they are remarkable for being more warmly clad than any other nation.' (a)

Rigby's theory did not satisfy Leslie whose own discussion, published one year before Crawford's book, dealt with all the contemporary theories in a careful and scrupulous manner. At this stage it made little difference to your theories whether you regarded heat or phlogiston as substances, or not. It was believed that both heat and phlogiston were evolved when vegetable and animal substances were burnt, and Leslie himself ascribes the cause of animal heat to the fact that the blood is 'burnt' in the course of the circulation. He attempts to show that blood contains phlogiston, which is evolved by the action of the blood vessels, and that this evolution is always accompanied by heat. He quotes the same passage from Franklin as Rigby, underlining the words 'that quicken the motion of the blood', and uses this as evidence that Franklin was 'of the opinion . . . that the heat of living animals is produced by the evolution of the phlogistic fluid'. One wonders whether Leslie had ever met Rigby; if so, their conflicting claims to the support of Dr. Franklin would have been worth hearing. Leslie says too that Dr. Mortimer anticipated Franklin's view by several years; he argued that

'. . . the elements of fire lay hid or dormant in bodies and that the AIR, which most substances both solid and liquid contain, being set at liberty by its elasticity excites into motion the latent particles of fire and generates heat, and therefore, as the animal fluid not only contains a large proportion of the phosphoric principle or sulphur in a quiescent state at least, he concluded that the generation of heat in

(a) 80, p. 79.

the vital frame was the necessary consequence of the particles of phosphorus and air coming into contact; and this he supposed was affected by means of the circulation.' (a)

Leslie endorses Mortimer's hypothesis as to the presence of phlogiston and phosphorus in the blood, but is not convinced that animal fluids contain air. He was a thorough phlogistonian, and though he knew of Lavoisier's work on mercury calx, he was not persuaded by it. Since according to his own theory, heat is produced in every part of the body as the blood circulates through the capillaries, it was not open to the objections which were made against Black's theory, to which we must now turn.

This fourth view is the one which, if we wish to trace the evolutionary descent of our modern ideas, must be considered as one of their common ancestors. It became widely known as 'Black's Theory of Animal Heat', and was still being considered seriously as late as 1843. Black certainly never published this theory, and there is no reference to it in his lectures as edited and published by Robison. But in his introduction Robison does refer to the idea.

'Dr. Black discovered that the breathing of an animal changes common air to fixed air; that this change is accompanied by the emission of heat; which emission seems to be the principal source of the heat generated in the bodies of all breathing animals.' (b)

He adds also:

'He conceived the accretion of solid matter as a source of a part at least of the warmth of animals.' (c)

Black's theory was, however, widely known and discussed. He discovered that

'. . . carbonaceous matter is separated and thrown off from the blood in the lungs in the act of respiration. In proportion to the quantity of blood which this (expired) air contains it is deficient in its due proportion of oxygene gas, a part of it having been changed into carbone gas.'

(a) 58, p. 94. (b) 13, p. liii. (c) 13, p. xxxviii.

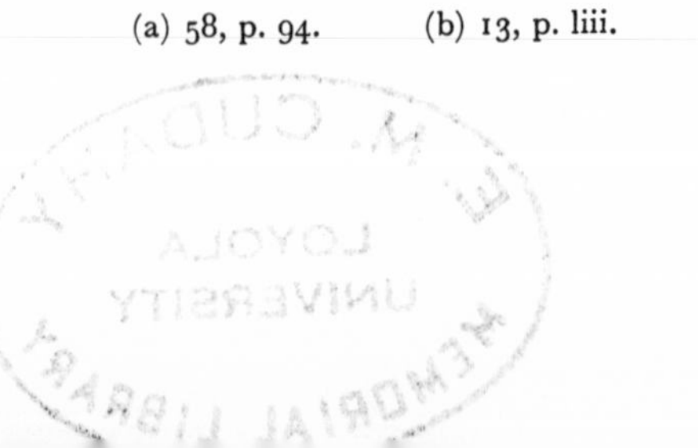

And he concluded from this observation that

> 'Animal heat depends on the state of respiration; that it is all generated in the lungs by the action of the air upon the principle of inflammability, in a manner little dissimilar to what occurs in actual inflammation; and that it is thence diffused by means of the circulation over the rest of the vital system.' (a)

Black's theory is accordingly a combustion theory: respiration is a kind of burning, and this is the source of animal heat. There was obviously a striking connexion between breathing and animal heat, because the speed with which the 'principle of inflammability' was separated by respiration seemed to be related to the degree of heat peculiar to each animal: birds, for example, breathe *faster* than mammals and maintain a higher temperature. It was also known that no heat was generated in the embryo until breathing was established.

Three objections to Black's theory were popular. The account I shall give of these objections is taken from Leslie's book, and is authoritative. Black's theory, he says,

> '. . . though never published is well known to us who have attended his lectures. Besides, I once had the satisfaction of a private conversation with the doctor on the subject, in which he explained to me his sentiments at such length, that I have every reason to believe that they are here delivered with sufficient accuracy and perspicuity.' (b)

According to some scientists, the fact that there was a correspondence between the phenomena of respiration and animal heat was not proof that there must be a causal connexion, because

> 'Certain animals have no organs of respiration and those fish which are destitute of gills appear to be warmer than the ordinary temperature of the medium.'

Evidently, breathing is not the whole story. According to the second objection,

> 'The Vital fluid rather than acquiring heat in passing through the lungs gives up what it has received during the course of circulation

(a) 58, p. 75. (b) 58, p. 89.

to the atmospherical air. Were blood heated in the lungs, we should certainly need less of their function in warm than in a cold atmosphere, but we are taught by experience that when the air is extremely hot and we wish to be cooled we breathe full and quick.' (a)

But the most forceful objection arose from the question of the temperature gradient. There was no proof that heat was carried by the arteries; if it was carried there should, as was clear by that time, be a noticeable heat gradient between the lungs and the heart; and lastly, the temperature of the lungs would be so much greater than the rest of the body that they would be unable to function properly. Here the wheel has come round again to the old difficulties of a central fire. Something must 'temper' this heat.

So far as we know, Black did not try to meet these objections. In one way this is surprising, since it was by applying Black's own theory of specific heats that Crawford was to answer the greatest of these objections. Black did however refer to Crawford's theory of animal heat in his lectures, implying that it was his, Black's work on specific heat and latent heat, that had given Crawford the basic idea.

Around 1770, therefore, there were several opinions about the cause of animal heat. Each of these, according to present-day views, was right in some respect. Haller and Leslie had the location of heat-production most nearly correct; Rigby was right in assuming that the food of animals contained heat in a bound form; Black showed that respiration was a factor in the production of animal heat. But the connexion between respiration and digestion had not been made. Nor could it be, so long as respiration was thought of as mere breathing, and an accurate chemical analysis of the blood fluids and gases had yet to be made. Thus none of the theories was completely satisfactory, and all had one feature in common; they lacked quantitative evidence in their favour; moreover some of the theorists probably did not even think that this kind of evidence was valid or necessary.

The unsatisfactory nature of these theories, which led many

(a) 58, pp. 75–90.

people to adopt an agnostic attitude, is summed up by Wriseberg[1].

'Although it be easier to show that animal heat is not produced properly from the friction of the fluids upon the vessels, than to substitute a more satisfactory and probable explication, my own reading and experience will not allow me to favour this theory. On the contrary I am persuaded, that the various states and quantity of animal power in the nervous system, the air's quality and accessibility, according to the freedom of their respiration, to our lungs, the motion of the muscles in their different degrees of increase and excitement, conjunctly with the progressive motion of the blood, not only produce the vital heat, but disperse it, when once produced, through all the body, in proportions which vary according to the greater or lesser supply of vessels in the different parts.' (a)

Hunter too was not at all satisfied, and remained a complete agnostic.

'The heat in the animal body . . . has been commonly considered as depending principally on the blood. As I shall have occasion to take notice of the increased heat of inflamed parts, it might be expected that I should endeavour to explain this principle in the history of the blood. I profess, however, not thoroughly to understand it, and the theories hitherto brought do not in the least satisfy me, as I think that none of them accord perfectly with every circumstance observable in these cases.' (b)

The first comprehensive, quantitative theories of the cause of animal heat were those of Adair Crawford (1779) and Antoine Laurent Lavoisier (1783). Lavoisier established on an exact basis what Black and others had already hinted at – the analogy between respiration and combustion. But he gave no answer to the objections already raised against Black's theory that combustion in the lungs was the source of animal heat; he did not explain how without any obvious temperature gradient the heat of combustion could become the heat of the whole animal system. The great attraction of Crawford's theory, to which we will turn later in this chapter,

[1] Wriseberg had supplied the notes to the translation of Haller's *Primae lineae Physiologicae*.

(a) 86, p. 106 footnote. (b) 47, p. 16 (*see also* 46, p. 282).

lay in the fact that, by an extremely ingenious application of Black's doctrine of specific heats, it appeared to meet these objections.

The comparison of respiration with combustion was an essential step in our story. But it was one which in certain respects was to prove misleading, since it focused on the lungs attention which eventually had to be transferred to the individual cells of the body. So long as the lungs were considered to be the site of actual combustion, rather than of mere gaseous exchange, the mechanism of heat-transfer remained a major problem. But, at the time that we are now discussing, the cell theory was scarcely imagined, and scientists were only beginning to suspect the extent to which arterial blood is capable of absorbing oxygen. For the moment, Lavoisier's work represented a real step forward.

Black, of course, had not been the first to suggest that respiration was a sort of combustion. The first seriously-argued hypothesis to this effect had been put forward by Mayow. He had suggested that the functions of the lungs was not to cool the blood but to form heat: he thought that the 'nitro-aerial spirits' in the air were absorbed and, mixing with the sulphureous particles of the blood, started up a kind of fermentation by which heat was produced.

> 'It is certain that animals in breathing draw from the air certain volatile particles which are also elastic, so that there should be no doubt now that something in the air absolutely necessary for life enters the animal by respiration . . . and indeed we must believe that the animals and fire remove particles of the same kind from the air.' (a)

Hooke had a similar idea.

> 'Animal heat may be caused by the uniting of the volatile salts of the air with the blood in the lungs which is done by a kind of corrosion or fermentation which to me I confess seems somewhat more probable.' (b)

Yet it was not until the end of the eighteenth century, with the work of Lavoisier, that the connexion between breathing

(a) 68, p. 300–6: *see* chapter *De Respiratione.* (b) 45, p. 50.

and the vital processes became firmly established. Lavoisier's contributions to the biological sciences were as great as his better-known contributions to chemistry. In both cases the key point was his emphasis on quantitative methods of study. The scruples which other scientists felt about extending to living things ideas and methods of experiment developed in physics and chemistry, never seemed to have worried him. He was prepared to apply directly to the study of animals those same quantitative methods that had proved so useful to him elsewhere. He never expressed himself on the question of whether there was any need to justify this step. The success of his experiments was presumably all the proof of the suitability of his methods that he required.

The story of his contributions to the problem of animal heat properly begins with his work on oxygen. This gas had been isolated about 1771 by Scheele, who gave it the name of 'fire air'. Priestley prepared oxygen from the calx of mercury in 1774; two months later he met Lavoisier in France and told him of the gas, which he called 'dephlogisticated air'. Priestley had noticed that it supported the breathing not only of a mouse but also of himself. His own view of respiration, or breathing (the two words were still being used synonymously), was that animals gave out phlogiston, and that breathing could continue only for so long as the atmospheric air remained unsaturated with this substance. Once saturation had occurred, no more phlogiston could be taken up from the animal and it eventually died.(a) Lavoisier repeated Priestley's work on the mercury calx during the latter half of 1774 and early part of 1775; he read his paper on this work to the Academy on 26 April, in that year. His simple but striking experiments, which are now classic, showed

'. . . that air which has served for the calcination of metals is, as we have already seen, nothing but the mephitic residuum of atmospheric air, the highly respirable part of which has combined with the mercury, during the calcination: and the air which has served the purposes of respiration, when deprived of fixed air, is exactly the same: and, in fact having combined, with the latter residuum, about

(a) 77, vol. ii, p. 44.

one quarter of its bulk of dephlogisticated air, extracted from the calx of mercury, I re-established it in its former state, and rendered it equally fit for respiration, combustion etc.: as common air, by the same methods as that I pursued with air vitiated by the calcination of mercury.'(a)

In a paper drafted in 1777 and published in 1780 he takes the question of respiration further.

'The result of these experiments is that to restore air that has been vitiated by respiration to the state of common respirable air, two effects must be produced. First, to deprive it of the fixed air it contains, by means of quicklime or caustic alkali: secondly, to restore to it a quantity of highly respirable or dephlogisticated air, equal to that which it has lost. Respiration therefore acts inversely as these two effects and I find myself in this respect led to two consequences equally probable, and between which my present experience does not enable me to pronounce . . .

'The first of these opinions is supported by an experiment . . . that dephlogisticated air may be wholly converted into fixed air by the addition of powdered charcoal: and, . . . it is possible . . . that respiration may possess the same property and that dephlogisticated air, when taken into the lungs, is thrown out again as fixed air . . . Does it not follow, from all these facts, that this pure species of air has the property of combining with the blood and that this combination produced constitutes its red colour? But whichever of these two opinions we embrace, whether that the respirable portion of the air combines with the blood, or that it is changed into fixed air by passing through the lungs: or, lastly, as I am inclined to believe, that both these effects take place in the act of respiration, we may, from facts alone, consider as proved:

'First, that respiration acts only on a portion of pure or dephlogisticated air, contained in the atmosphere; that the residuum or mephitic part is merely a passive medium which enters the lungs and departs from them in nearly the same state, without change or alteration . . .

'Thirdly, that in a like manner [to his second point on calcination] if an animal be confined in a given quantity of air, it will perish as soon as it has absorbed, or converted into fixed air, the major part of the respirable portion of the air and the remainder is reduced to a mephitic state.

(a) 50, *see* Hall, p. 19.

'Fourthly, that that species of mephitic air which remains after calcination of metals, is in no wise different . . . from that remaining after the respiration of animals . . .' (a)

But it was in a memoir on combustion presented to the Academy on 5 September, 1775, that his suggestions about the source of animal heat were first made.

'I showed in a memoir which I read at the public meeting last Easter that pure air, after having entered the lungs leaves in part as fixed air or the acid of chalk. Pure air, in passing through the lungs undergoes then a decomposition analogous to that that takes place in the combustion of charcoal. Now in the combustion of charcoal the matter of fire is evolved whence the matter of fire should likewise be evolved in the lungs in the interval between inhalation and exhalation, which distributed with the blood throughout the animal oeconomy, maintains a constant heat of 32½ degrees Reamur. This idea will appear to be hazarded at first glance, but before it be rejected or condemned I beg you to consider that it is founded on two certain and incontestable facts, namely, on the decomposition of the air in the lungs and on the evolution of the matter of fire which accompanies all decompositions of pure air, that is to say all changes of pure air to the state of fixed air. But that which further confirms . . . is that only those animals in Nature which respire habitually are warm blooded and that there is a constant relation between the warmth of an animal and the quantity of air entering, or at least converted, into fixed air in its lungs.' (b)

Having established an analogy between combustion and the respiration of animals on a qualitative basis Lavoisier then tried to produce quantitative proof. In 1783 appeared his famous paper written jointly with Laplace. (c) This paper is worth careful analysis. It is divided into four parts. In the first part the authors give details of a new method of measuring heat. They make clear what they mean when they use the term 'free heat', 'capacity for heat' or 'specific heat of bodies' and briefly refer to the opposing theories concerning the nature of heat which were held by the physicists of their day. They rightly point out that the fundamental principle on which their experiments are conducted – i.e. that the quantity of free heat remains the same when two bodies are

(a) 50, Hall, p. 198. (b) 52, Klickstein, p. 172. (c) 54.

mixed together – is equally plausible whatever the view adopted about the nature of heat.

'During the simple mixing of substances the quantity of heat always remains the same . . . if heat is a fluid which tends to remain in equilibrium, or, if it is only a motive force which results in internal movements in substances, nevertheless it always obeys this principle of conservation. The conservation of free heat, in the simple mixing of substances, is always independent of all hypotheses as to its nature: this is generally agreed by physicists and we adopt this attitude in our researches.' (a)

In their experiments the quantity of heat produced by the combustion of materials or the respiration of animals was found by measuring the amount of ice melted in an ice calorimeter; and most of the first section of the paper is taken up with descriptions of their methods. The apparatus used consisted of three chambers, the innermost of which contained the object whose heat-production was to be measured, while the middle and outer chambers were filled with crushed ice. The ice from the middle chamber on melting ran down through a grill and sieve, and was collected beneath the apparatus: the ice in the outer chamber was intended to prevent heat from the external air affecting the results of the experiment.

The second section of the paper reports the results of some of their experiments, and lists the specific heats that they had calculated for water, mercury, and a wide range of other substances. They went on to compare the quantity of ice melted by the heat produced by a guinea pig in ten hours with that melted by the heat of combustion of various substances.

The simplest account of this part of the experiment is given by McKie in his biography of Lavoisier, and for the sake of brevity I shall quote him.

'The experimenter (1) put a guinea pig into the ice calorimeter for ten hours and measured the resulting amount of ice melted, which, of course, measured the heat given out by the animal during that time;

(a) 54, p. 12.

the heat evolved melted 13 ounces of ice. Then (2) they determined, in a separate apparatus, the amount of fixed air produced by the combustion of a weighed quantity of charcoal; and (3) they measured, in the ice calorimeter, the heat produced by the combustion of another weighed quantity of charcoal. From (2) and (3) they could calculate the heat produced in the formation of any quantity of fixed air.

'Next, (4), they determined how much fixed air was produced by the respiration of a guinea pig in ten hours, the interval during which its loss of heat had been measured in (1). Calculation from (2) and (3) then gave the quantity of heat evolved in the formation of this amount (4) of fixed air: it corresponded to the melting of 10 ounces of ice.' (a)

The fourth part of the paper is concerned with the analogy between combustion and respiration, and in it Lavoisier and Laplace explain the results of their experiments. To begin with they recapitulate the current views on the nature of the gaseous exchange in respiration, pointing out once more that air which is fit for respiration forms only about one-quarter of the atmospheric air. This air is either absorbed or altered, or converted into fixed air by the addition of a principle which, in order to avoid the necessity of discussing its nature, they call the 'base' of fixed air. Respiration, they conclude, is not just a physical, but a chemical process

'. . . the air does not act simply as a mechanical force but as an agency of new combinations.' (b)

They then give an account of their methods for calculating the quantity of fixed air produced in respiration and combustion, and finally evaluate their results.

'It was previously seen that in the combustion of carbon, the formation of an ounce of fixed air can melt 26·692 ounces of ice; on the basis of this result it is found that the formation of 224 grains of fixed air must melt 10·38 ounces. This amount of melted ice consequently represents the heat produced by the respiration of a guinea pig during ten hours.

'In the experiment on animal heat of a guinea pig, this animal emerged from the apparatus with nearly the same heat with which it entered, for it is known that the internal heat of an animal is always nearly constant. Without the constant renewal of its heat, all the heat

(a) 61, p. 143 (Schuman Edition, New York). (b) 54, p. 58.

which it had at first would have been gradually dissipated, and we should have found it cold upon taking it out of the apparatus, like all the inanimate objects we have used in our experiments. But the animal's vital functions continually restore to it the heat which it gives off to its environment, and which is in our experiment diffused into the inner ice of which it melted 13 ounces in ten hours. This amount of melted ice thus represents approximately the amount of heat renewed during this time interval by the vital functions of the guinea pig. Perhaps an ounce or two should be subtracted, or maybe more, on account of the fact that the extremities of the body of the animal were chilled in the apparatus, although the interior of the body retained nearly the same temperature; furthermore the moisture which its internal heat had melted evaporated a small amount of ice as it cooled, adding to the water draining out of the apparatus.

'On subtracting 2½ ounces from this quantity of ice, one obtains the amount melted by the effect of respiration of the animal upon the air. Now if one considers the inevitable errors in these experiments and in the factors which were the starting points for our calculations, it will be seen that it is not possible to hope for a more perfect agreement between the results. Thus the heat which is liberated in the transformation of pure to fixed air by respiration may be regarded as the principal cause of the conservation of animal heat . . .

'Respiration is therefore a combustion, very slow to be sure, but perfectly similar to that of carbon. It occurs in the interior of the lungs, without the liberation of any perceptible light because the fire as it is freed is absorbed by the humidity of these organs. The heat developed by this combustion is transferred to the blood which passes through the lungs, and thence is transmitted throughout the animal system. Thus the air we breathe serves two purposes equally necessary for our preservation: it removes from the blood the base of fixed air . . .; and the heat which this combination releases in the lungs replaces the constant loss of heat into the atmosphere . . .' (a)

A later paper by Lavoisier and Seguin adds two more suggestions. The first is that a given volume of carbonic acid gas contains less bound heat than the same volume of oxygen, so that free caloric is disengaged during the combustion of carbon. The second is that, during respiration, not only is some oxygen taken up by the blood, but a small amount of it combines with hydrogen contained in the blood

(a) 54, *see* Gabriel, p. 92.

to form water. They also hint in passing that the difference between the specific heats of arterial and venous blood may result in heat production

> '. . . the vital air loses a part of its specific heat; the effect of respiration is to remove from the blood a portion of carbon and hydrogen and give up in its place its specific heat which, during the circulation, is distributed . . .' (a)

But this allusion is not clear enough to show whether Lavoisier had in mind an idea similar to Crawford's[1] (which we shall discuss next). At any rate, Lavoisier was obviously aware of the theoretical difficulty involved in explaining why, if a combustion capable of producing all the heat required by an animal is taking place in the lungs, they do not in consequence burn up.

In Britain, Dr. Adair Crawford was doing similar work. In 1779, he reported the results of his experiments in *Experiments and Observations on Animal Heat* (b), and this book was very favourably received both at home and on the Continent. Crawford was a physician at St. Thomas's Hospital, London, and later Professor of Chemistry at the Royal Military Academy, Woolwich. He visited Scotland in 1776 and in 1777 began his experiments on animal heat in Glasgow, Black's old university.

Crawford does not specify the extent of his debt to Black but it must have been considerable. In most of his experimental work he used those quantitative physical methods which are usually associated with Black's name. He was

[1] In Article 4 of their paper, Lavoisier and Laplace refer to Crawford's experiments, from the first edition of his book (1779), and show the essential difference between his theories and theirs:

> '. . . this [difference] is that Mr. Lavoisier believes that the heat released in these two processes [respiration and combustion] is found combined in pure air, the gaseous state of this fluid being due to the expansive force of the heat combined with it; on the other hand, following Mr. Crawford, the matter of heat is free in pure air, it is released because on reaction, the pure air loses a large part of its specific heat.' (c)

They mention the delicate nature of Crawford's experiments and leave the discussion of his work by stating that they were only estimating the quality of heat produced in respiration and combustion without examining the actual source of that heat.

(a) 55, p. 36. (b) 23. (c) 54, p. 59.

further indebted to him for the ideas of specific and latent heats, on which his own theory of animal heat was based. Yet, as McKie and Heathcote point out, from the first edition of Crawford's work

> 'The unwary reader . . . might well imagine . . . that he was being introduced, if not to one of the founders of the quantitative science of heat, at least to one of its extensive improvers.' (a)

Crawford indicated only that the doctrines of specific and latent heat had been taught in Edinburgh by Black and Irvine for 'several years': he mentioned Black's doctrine of latent heat only vaguely, though in the second edition he did give Black credit for its discovery. We cannot know whether or not Crawford's experiments were specifically designed to meet the objections against Black's theory of animal heat. Like Lavoisier, Crawford gives no indication that Black ever had such a theory. This omission is more surprising in Crawford's case than in Lavoisier's. Black's habit of leaving his work unpublished was well known, but since Crawford had visited Edinburgh and talked to Irvine – and according to some accounts, had even worked with Black – it would be very surprising if he had not known of the latter's ideas. His omissions in this respect particularly surprised his contemporaries, since (as Bostock puts it) he was distinguished for his 'candour and liberality'.

The following account of Crawford's theory is taken from the second edition of his book published in 1788. In the nine years between the appearance of the first and second editions he had repeated many of his experiments, since much early criticism of his work had been concerned with possible experimental errors, and he was determined to eliminate these as far as he could.

He begins by defining 'absolute heat' – 'which expresses, in the abstract, that power or element which, when it is present to a certain degree, incites in all animals the sensation of heat'. He uses the term relative heat as expressing 'the same power considered as having a relation to the *effects* by which it is

(a) 62, p. 38.

known and measured. There are, he says, three effects by which heat is known. There is the sensation which it excites in animals and which can be felt as 'sensible heat'; there is its effect on instruments specifically designed to measure its 'temperature of heat in bodies', and lastly, there is 'comparative heat'. This last is a ratio of the amount of absolute heat in one body 'considered relative to another'.(a) We are here concerned more with Crawford's theories than with the actual details of his experiments, but it is interesting to notice some of the corrections that he made, since they reveal his care and skill as an experimenter. His first experiments were devised to determine the comparative quantities of absolute heat in bodies by mixing them together and observing the changes which are produced in their sensible heats. This was a continuation of the calorimetric methods devised by Black, and Crawford, as well, was aware of the need for a cooling correction when employing the method of mixtures. Crawford points out the inaccuracies that must be expected when taking water as a standard; since it transmits heat very slowly, and as 'great agitation' is needed to mix another liquid into it, there is considerable danger of loss of heat.

In the first series of experiments he calculates the comparative quantities of absolute heat in certain vegetable and animal substances, showing that milk, flesh and vegetables contain comparatively less absolute heat than water, and water less than arterial blood. On account of the tendency revealed by these experiments for arterial blood to accumulate heat he suspects that, during respiration, arterial blood absorbs heat from the air. He finds confirmation for this in three facts: first, in the evidence, also mentioned by Lavoisier, that only those animals which have lungs and continually 'inspire the fresh air' have the power of keeping themselves at a temperature above that of their surroundings; second, in the fact that warm-blooded animals have the most extensive respiratory organs; and finally, in the observation that animal heat is increased by exercise and by anything else that accelerates breathing.

(a) 23, p. 3.

Crawford now sets out a series of propositions, which are followed by experiments designed to prove them.

'*Proposition 1.* The quantity of absolute heat contained in pure air, is diminished by the change which it undergoes in the lungs of animals and the quantity of heat in any kind of air that is fit for respiration, is nearly proportional to its power in supporting animal life.' (a)

He considers first the nature of the changes which occur during respiration and agrees with Lavoisier's analysis, except that he retains the phlogistonian terminology.

'And this alteration consists of the conversion of pure into fixed air, by the union which the former contracts with inflammable air or with its basis in the process of respiration . . . It is well known that the blood undergoes a remarkable change of colour when circulating in a living animal, for the vivid arterial blood, in its passage through the capillaries to the venous system, acquires a deep and livid hue and again resumes its bright and florid colour in the lungs.' (b)

He notices that Priestley produced this change of colour in blood outside the body by exposing it to pure and inflammable air, and that the colour changes took place even though a thin bladder was interposed between the air and the blood. He also quotes the experience of his friend Dr. Hamilton, who introduced inflammable air into the jugular vein of a cat in order to show that this change can occur inside the living body as well. Crawford is well aware that there are two species of inflammable air and that the one used by Hamilton, was produced by the action of vitriolic acid on iron filings, was not the same as that which is obtained from animal substances; but he adds, inconsequentially,

'. . . it cannot be doubted that the same effect would be produced by exposing blood to that species of inflammable air which is obtained from animal substances.' (c)

Crawford concludes that the 'inflammable principle' is absorbed in the capillaries and lost again when the blood recovers its florid colour in the lungs. Quoting the experiments

(a) 23, p. 144. (b) 23, p. 147. (c) 23, p. 150.

of Priestley and Cavendish, he argues that pure air is received into the lungs, where it combines with the inflammable principle to form, in part, aqueous vapour and, in part, fixed air.

Crawford sets out to prove his first proposition by attempting to show that pure air contains a greater quantity of absolute heat than the fixed air and aqueous vapour which are exhaled from the lungs. These experiments were evidently very difficult to perform with any degree of accuracy, and of all the things in his book they best reveal his ingenuity. The differences between the first and second editions show what efforts he made to increase the accuracy of his methods during the years between their publication. He lists the difficulties that confront anyone who wishes to determine the comparative heats of different kinds of air. These arise, he says,

'. . . partly from the fugitive nature of heat, partly from the rarity of the permanently elastic fluids, in consequence of which not more than 15 or 20 grains can conveniently be employed in an experiment, and partly from the changes they are liable to undergo, with respect to purity and the quantities of moisture which they contain.' (a)

He concludes that the amount of heat given out by the air is so small that accurate thermometric measurements are very difficult.

As before, he employs the method of mixtures. He first tries to force atmospheric air which had been heated in a bladder over cold iron filings; but he notes that the iron always 'contracted moisture', which he presumes to have come from the bladder, and which he can never entirely eliminate. He then heats the air in Florence flasks and immerses these in water of known temperature. This method too was subject to a number of inaccuracies, which arose mostly from the difficulty of sinking the flask to the same depth every time. He also tries using wine thermometers – which provide a large distance between two consecutive points on the scale – as well as air thermometers. The wine thermometers he finds 'defective in point of sensibility'; the air 'though very

(a) 23, p. 157.

sensible, being deficient in point of accuracy'. Having then listed eight 'circumstances which require attention' when experiments for this determination are made, he describes the apparatus which he designed to take account of them.

This consisted of two brass cylinders of identical weight and capacity attached to a central handle. Each could be filled with a different species of air and carried a thermometer. The cylinders were lowered simultaneously into two tinned vessels of equal weight and dimensions, filled with either water or oil at room temperature, and lagged throughout the experiments with thick flannel. The brass cylinders were filled with the two gases under consideration, heated in two more vessels placed in a water bath – great care being taken that no moisture should come into contact with the gases – and then immersed in the tin vessels. The experiment then proceeded along standard lines, enabling Crawford to measure the different comparative heats of the gases relative to water or oil. He found that, comparatively, pure air contains more heat than atmospheric air, and that dephlogisticated air contains a greater quantity of absolute heat than fixed air, phlogisticated air or aqueous vapour. He concludes:

'Hence the quantity of heat in a mixture of pure and phlogisticated air, will be increased by augmenting the proportion of the former, and diminished by augmenting that of the latter. Agreeably to this we find that the comparative heat of pure air is to that of atmospherical, which consists of a ¼ pure and ¾ phlogisticated air, as 2·2 to 1 . . . The quantity of *pure air* in a given portion of the purest dephlogisticated air is four times as great as that contained in an equal bulk of atmospherical; and that the comparative heat of the former is to that of the latter only as 2·2 to 1. The reason is obvious: phlogisticated air which is one of the constituent parts of the atmosphere, contains a considerable quantity of absolute heat . . . If from the heat of a given quantity of atmospherical air, we deduct the heat of the phlogisticated air, which forms one of its constituent parts, and which is not altered by respiration, the quantity of heat in the remainder will be to that contained in a quantity of pure air, equal in bulk to the atmospherical, as 1 to 4 nearly. And as Dr. Priestley . . . has proved, that the power of the latter, in supporting animal life, is nearly five times as great as that of the former . . . *There is . . . sufficent evidence for concluding that the purer part of the atmospherical air has its capacity for heat diminished by the change which it undergoes in the lungs of animals.*'[1] (a)

This last sentence is to be crucial for his theory of animal heat.

Crawford's second proposition runs:

'The blood which passes from the lungs to the heart by the pulmonary vein, contains more absolute heat than that which passes from the heart to the lungs by the pulmonary artery.' (b)

This he proved by a series of experiments using the same method of mixtures, mixing the two kinds of blood with known weights of water; and he found that the ratio of the heat of the venous to that of arterial blood was nearly 10 to 11½.

His third proposition reads:

'The comparative quantities of heat in bodies, supposed to contain phlogiston, are increased by the changes which they undergo on the processes of calcination and combustion.' (c)

At the onset he refers to Lavoisier's objections against the phlogiston theory and gives his reasons for rejecting them. He next proves, again by the method of mixtures, that when metals are calcined their quantities of absolute heat are increased. However, because of the loss in weight which occurs when substances are burnt, he cannot adopt the same technique for determining to what extent the absolute heat is increased under these circumstances, and he has to argue indirectly.

'We may conclude in general that when substances which resist the force of fire are united to pure air, aerial acid or water, the capacity of the compound is greater than that of the fixed substance.' (d)

Though his experiments showed that wood and pit-coal have more absolute heat than their ashes, so that the heat-capacity of these substances must be diminished by combustion, he nevertheless reasons

'. . . it is manifest that the sum of the capacities of the principles into which those substances are resolved by that process, is increased. For fixed bodies, as was before shown, have a less capacity for heat

[1] My italics. (a) 23, p. 47. (b) 23, p. 273.
(c) 23, p. 279. (d) 23, p. 299.

than pure air, aerial acid, or water. When these substances enter into chemical union with each other, it is very probable that capacities of the fixed parts are increased (a) . . . *It follows that when a body is deprived of its power of supporting flame, by the process of combustion, it absorbs a quantity of absolute heat, and when, by a contrary process, it again recovers its inflammability, an equal quantity of heat is detached.*'[1] (b)

This was the only point in his analysis at which he was unable to provide direct experimental proof.

Crawford's fourth proposition comprises three points.

'When an animal is placed in a warm medium, the colour of the venous blood approaches more nearly to that of the arterial than when it is placed in a cold medium; the quantity of respirable air which it phlogisticates, in a given time, in the former instance, is less than that which it pholgisticates, during an equal space of time, in the latter; and the quantity of heat produced when a given portion of pure air is altered by the respiration of an animal, is nearly equal to that which is produced when the same quantity of air is altered by the burning of wax or charcoal.' (c)

The last part of this proposition is of course what Lavoisier was trying to prove.

He easily proves the first point by keeping a dog in water at various temperatures with only its head uncovered, and drawing small quantities of blood from an artery and a contiguous vein. The colder the water in which the animal was immersed, the darker the venous blood became. These experiments confirmed, in Crawford's opinion, a suggestion made by his friend, Mr. Wilson.

'Admitting that the sensible heat of animals depends upon the separation of absolute heat from the blood by means of its union with the phlogistic principle in the minute vessels, may there not be a certain temperature at which that fluid is no longer capable of combining with oxygen and at which it must of course cease to give off heat?'

He then experiments on guinea pigs, showing that the air 'expired from the lungs is more phlogisticated in a cold than in a warm medium'. The animals were kept in jars containing

[1] My italics. (a) 23, p. 303.
(b) 23, p. 305. (c) 23, p. 307. (d) 23, p. 311.

about five pints of common air, and placed in water at different temperatures for thirty-six minutes. After this time the air in the jar was examined to determine its degree of purity.

Crawford's last series of experiments were similar in aim to those of Lavoisier and Laplace, but his technique was different; the quantity of heat given off by a specimen was determined by warming water rather than by melting ice. His results showed an even closer correspondence between combustion and respiration than those of the two Frenchmen. He found that 100-ounce measures of pure air, altered by the combustion of wax, by the combustion of charcoal, and by the respiration of a guinea pig, communicated to the same amount of water 21°, 19·3° and 17·3° of heat respectively. (Here Crawford is expressing quantity of heat as degrees of temperature communicated to water: Lavoisier and Laplace expressed it in terms of the weight of ice melted.) Crawford comments on the results of Lavoisier and Laplace, who found the ratio of the heat formed by the combustion of charcoal and the breathing of a guinea pig to be 10·3° to 13°. He accounts for this discrepancy by pointing out that their guinea pig was placed in a temperature of 60° when the amount of air altered by respiration was measured, and in a temperature of 32° or 33° F. when the heat it produced was determined. He had proved that an animal at a lower temperature phlogisticated a greater quantity of air than it would at a warmer temperature, so his results could not be expected to agree with theirs. He notices that the cause of this inaccuracy was fully appreciated by the Frenchmen. The discrepancy in his own results he explains as follows

'. . . in the first instance, a considerable part of the air is converted into water, and more heat is produced by the conversion of pure air into that fluid than into fixed air. In the last instance, a part of the heat is carried off by the insensible perspiration.' (a)

In the light of all these experiments Crawford is in a position to offer a complete account of the mode of production and transmission of heat in animals.

(a) 23, p. 352.

'It has been proved that the air which is received into the lungs of animals contains more absolute heat than that which is exhaled . . . It has been shown that nearly one sixth of the pure air which is altered by the process of respiration is converted into aqueous vapour and the remaining five-sixths into fixed air; . . . and that when the quantities of matter in these fluids are equal, the comparative heats are as 4 to 1; from which it appears that, the quantity of matter in fixed air being to that in pure air as 5 to 4, the quantity of heat in the latter will be to that in the former as 3 to 1 nearly; and it has been proved that this likewise is the heat of pure air to that of aqueous vapour.

'Since, therefore, the fixed air and aqueous vapour which are exhaled by expiration are found to contain only one third part of the heat which was contained in the purer part of the atmospherical air, previously to inspiration, it follows that the latter must necessarily give off a considerable portion of its absolute heat in the lungs.

'It has moreover been shown that the comparative heat of florid arterial blood is to that of venous as 11½ to 10, and hence, as the blood which is returned by the pulmonary vein to the heart, has its quantity of absolute heat increased, it is evident that it must have acquired this heat in its passage through the lungs. We may conclude therefore that, in the process of respiration, a quantity of absolute heat is separated from the air and absorbed by the blood . . .

'From the foregoing experiments and observations it follows that animal heat depends upon a process resembling a chemical elective attraction. The pure air is received into the lungs containing a great quantity of elementary fire; the blood is returned from the extremities impregnated with the inflammable principle; the attraction of pure air to the latter principle is greater than that of the blood. This principle will therefore leave the blood to combine with the air; by this combination the air is obliged to deposit part of its elementary fire and as the capacity of the blood is at the same moment increased, it will instantly absorb that portion of fire which had been detached from the air. The arterial blood in its passage through the capillary vessels is again impregnated with the inflammable principle in consequence of which its capacity for heat is diminished. It will therefore in the course of circulation gradually give out the heat which it had received in the lungs and diffuse it all over the whole system.' (a)

(a) 23, p. 361.

To paraphrase the theory in twentieth-century terms; (i) the oxygen that an animal breathes in, is converted in the lungs into carbon dioxide and water vapour; (ii) the inspired oxygen contains more heat than the expired products of respiration; (iii) the balance of heat released in the conversion is taken up by the arterial blood without any increase in temperature; (iv) this is possible because the change from venous to arterial blood in the lungs is accompanied by an increase in the specific heat of the blood; and (v) the heat taken up by the arterial blood is released throughout the body when it is converted back into venous blood in the capillaries, and its specific heat is correspondingly reduced. Crawford's quantitative observations were consistent with this ingenious theory, which at one stroke overcame the gravest objections which had previously affected all combustion theories of animal heat. This theory, as Pritchard was later to remark, was one of 'the most splendid and elaborate . . . ever constructed in any subject in physiology.'(a)

When one compares the work of Lavoisier and Crawford, one is struck by the difference in their aims. Lavoisier, as he himself points out, was concerned to show only that a qualitative and quantitative analogy existed between respiration and combustion: this he believed to be true, not only because the same air – oxygen – was used up in both these processes; but also because the same amount of heat was produced. As we have seen, his experiment did not show an exact quantitative relationship between respiration and combustion. Nevertheless, the results were sufficiently striking, especially when taken *in conjunction* with his work on the nature of the gaseous exchanges which take place in the respiration and transpiration of animals. But Lavoisier was not concerned, as Crawford was, with the actual mechanism by which the heat was evolved. He was aware that this had to be accounted for; but he did not say whether, in his view, the heat came from the decomposition of the air, from the fixation of the carbonaceous base in the lungs, from the formation of water; or whether it was due to the different thermal capacities of

(a) 78, p. 110.

oxygen and carbonic acid gas, which led to a release of heat when new compounds were formed.

Although Crawford did not explain *in detail* the production of heat during combustion, he did show how the lungs and the blood would remain at a uniform temperature. The arterial blood, having, he believed, a greater capacity for heat than venous blood, became saturated in the lungs with the heat produced during the change of oxygen to carbonic acid gas. So the temperature of the blood and the lungs did not alter.

According to Lavoisier's theory, heat is set free in the lungs. He presumes that the constant temperature of all parts of the body is maintained by three factors: the speed of the circulation, by which the heat freed in the lungs is taken to all parts of the body; evaporation from the lungs, which removes part of the heat; and possibly also the increased capacities for heat which blood acquires when it is converted from the venous to the arterial state. In one sense Lavoisier held to the older view that the lungs cool the body – according to him by evaporation of the water produced – even though he also considered them to be the place where bodily heat was originally formed. Crawford, on the contrary, believed that heat was not actually set free until the blood had circulated as far as the capillaries, where the 'inflammable principle' was combined with it, thus reducing its heat-capacity and releasing sensible heat. This view had one further merit, which must impress us today; it transferred the evolution of heat to the capillary system, where it was in any case most necessary to counteract the loss of heat to the external surroundings. Quite apart from the fact that it explained the uniform temperature of the body, this insight was to become especially important with the introduction of the cell-theory fifty years later.

The experimental methods used by the two men are also worth comparison. In this respect, Lavoisier has the advantage. Crawford's determinations of specific heat depended on the method of mixtures, and the quantity of heat was estimated by the rise in the temperature of water; Lavoisier, on the other hand, measured the weight of ice melted. Lavoisier's experiments involving the latent heat of fusion of ice, can be made more accurately than Crawford's for two reasons: first, because balances are more accurate than thermometers, and second, because one does not need extensive cooling corrections. In any case, Crawford was attempting something far more difficult. To estimate the specific heats, or as he would have called it, the 'quantities of absolute heat in bodies', of solids was difficult enough; but to estimate those of gases was worse still; and to do this for bodies before and after combustion was almost impossible.

These men should be compared on two last points. Crawford, unlike Lavoisier, was a phlogistonian; although it cannot be said that his theory would have needed material alteration had he abandoned the phlogiston theory. In effect, the only difference between them was that where Lavoisier would write of 'the principle of oxygen', Crawford would write of 'dephlogisticated air'. Crawford had in any case a very moderate attitude towards phlogiston, accepting it as part of the chemical theory of his day without ascribing to it implausible and inconsistent properties. For a man like Leslie,(a) phlogiston was a stuff with pretty remarkable properties, on which his own theory of animal heat depended; it was even more remarkable in the hands of a man like Peart,(b) who rejected both Crawford's and Lavoisier's theories of animal heat, and argued that it arose from the action of two separate imponderables, Phlogiston and Aether.

Both Crawford and Lavoisier were aware of the suggestion that oxygen, instead of being burned in the lungs, might be absorbed by the blood; and of the two men Lavoisier was the more prepared to accept this as a reasonable possibility. It would complicate his theory, because it would presumably entail the shifting of the place of combustion away from the lungs, and instead distribute combustion throughout the body. However, he skirted round this difficulty by supposing that most of the oxygen was united to the carbonaceous base in the lungs, while *some* of it combined with hydrogen thrown out by the blood to form water, and a small volume only was absorbed

(a) 58. (b) 75.

by the blood.(a) Crawford found no decisive experiment to justify a belief in the absorption of dephlogisticated air, but he remarked that even if it were absorbed this fact would not affect his theory at all.(b) The crucial point in his explanation is the observed change in the specific heat of the blood; and whether this takes place through absorption of oxygen or for other reasons is immaterial.

Finally, we may compare the two men's ideas about the nature of heat. Lavoisier, after considering existing opinions on this subject, ultimately came to regard it as 'matter of fire', and coined for it the name of 'caloric'. Gas is for him, then, a 'principle' or 'base' combined with caloric. In his *Elements of Chemistry* (c) he sets out the difficulties involved in accounting for all the phenomena of heat without admitting it to be a real and material substance. He gives the name 'caloric' to 'the cause of heat', which he himself supposes to be an 'exquisitely elastic fluid'; but, he points out, we are not compelled to imagine this to be a real material substance,(d) and in the experiments which he did on animal heat its exact nature does not affect the issue at all. Crawford comes to a similar agnostic conclusion, and so does not trouble to discuss the nature of heat, saying only that it would be very difficult to 'reconcile many of the phenomena on the supposition that heat is [nothing more than] a quality'.(e) The nature of heat was not at this stage relevant to discussions of the source of animal heat. It was only later, with the introduction of the kinetic theory of matter and the development of the concept of energy, that the issue between the two views became important to physiology. But by that time, with the gradual recognition that oxygen was absorbed by the blood and the development of the cell theory, the focus of problem had shifted, and become centred in the tissues of the body.

Crawford's work made a tremendous impact, although in the long run it was Lavoisier's that proved to be the more fruitful. Crawford's theory of animal heat was almost too perfect and complete: it seemed to explain completely how,

(a) 55, p. 36. (b) 23, p. 155.
(c) 53. (d) 53, p. 5. (e) 23, p. 435.

given a combustion in the lungs, heat could be safely distributed throughout the animal's body and then released where it was most required. It was based on carefully designed experiments, and expressed in terms of chemical and physical ideas – combustion and specific heat – that were very much in fashion at the time. It was still receiving serious consideration sixty years after its first publication, and it was never directly refuted. Rather, it was eventually by-passed. And as methods of studying chemical and physical processes within the body developed, from the 1820s on, it gradually fell out of view.

3

The Methodological Situation around 1800: Bichat's Scruples

THE success of Crawford's and Lavoisier's work put physiologists in a quandary. It had been assumed that the activities of living things were quite different, both in character and cause, from those of inert things. Once a direct, quantitative analogy between the two realms had been established, the question became urgent – what constitutes the unique quality of living things? Though the terrain of debate has changed from century to century, this has always been the chief question in biology on which mechanists and vitalists divide: and it is instructive to study the form in which the issue was debated around 1800. For the crucial question now became, how far it was legitimate to apply the methods of investigation used by physicists and chemists to living organisms; and in particular, whether searching for exact mathematical relations between biochemical variables did not commit one to a rigidity in physiological explanation which was inconsistent with the variability and complexity obvious to any observer of animal behaviour. Compared with this, the thesis that organic substances are of a unique kind – through his refutation of which Wöhler is often supposed to have dealt a fatal blow to vitalism – was a secondary issue.

In considering how scientists' opinions were divided at this time, it is essential not to allow the unfavourable modern associations of the word vitalism to mislead one. Earlier vitalists, from Van Helmont up to Barthez, had certainly used their 'vital principles' as explanatory factors: the unique, constructive and integrated activities of organisms (they thought) could be accounted for only by supposing them to be manifestations of a single, unifying agency residing within each organism, and peculiar to it. (The exact names that this agency was given were unimportant.) But one could have scruples over the validity of arguing about the processes going on within organisms in a purely physico-chemical way, without going to this extreme of 'explanatory vitalism'. And much of the discomfort shown by clinical physiologists and others around 1800 springs from a far more modest and justifiable species of vitalism. The variability and complexity of animal behaviour were – for them – obvious facts, at a level of straightforward description. No theory which excluded them was worth serious consideration. Such men would not have denied that Lavoisier's experiments demonstrated something; what they did deny was that they were of serious relevance to the central problems of physiology.

Some analogy between combustion and respiration there might be, but it was so remote as to leave the essential parts of the story untold.

Accordingly, one must not impatiently sweep aside the scruples of men like Bichat, as physiologists from Magendie onwards have been tempted to do. In 1800, the extent to which physics and chemistry could provide explanations of physiological behaviour was extremely limited, so that, on the one hand, the hypothesis of a unique 'vital force' (such as Liebig was to favour) could not be summarily ruled out, and, on the other, to believe that physics and chemistry would eventually give the physiologist all the explanatory factors he needed was a complete act of faith. Meanwhile, at the level of behaviour, the points which the more modest 'descriptive vitalists' made about the uniqueness of life were (and still are) absolutely correct.

The methodological question remains. If we accept that the quantitative physics and chemistry of inert matter cannot yet tell us all we need to know in order to explain the behaviour of living things, can they nevertheless tell us something? Here Bichat's conclusions were to prove unfortunate. For, from the essential plasticity of physiological phenomena, he inferred an inexactness in the application of physico-chemical laws to them. We may not like this conclusion, but the quandary in which he found himself was undeniable. And just how one was to square determinism and mathematical exactitude at the biochemical level, with plasticity at the gross physiological level, was a question which remained unanswered until the time of Claude Bernard.

Let us consider the background to Bichat's work a little more closely. By 1800 the main characteristics which distinguish living organisms from non-living material had been identified and described. The unique activities of living organisms, such as irritability and growth, had been established and classified, mainly by medical men such as Haller and Hunter. Once this was done, the main question became, what are the sorts of terms in which the causes of these vital activities are to be explained? Clearly this was not an entirely new question; the earlier ideas of a unique vital substance, principle, or constituent, had provided and still continued to provide qualitative explanations of a sort. But now physical and chemical explanations joined the list of possible answers.

The experiments described in the previous chapter were possibly the first where an explanation of a specifically biological activity was offered in purely physico-chemical terms. The animals studied were part of a physico-chemical experiment, and direct quantitative measurements were made to demonstrate how the animal 'worked'. Both experimenters seem to have taken it for granted that they were justified in employing the procedures they did, and to that extent they were both 'methodological mechanists'. When Lavoisier[1] uses the phrase 'vital principle' it is never as the name for a hypothetical explanatory factor of a non-physico-chemical kind, but as a descriptive phrase just like 'animal machine'. And when in his last paper on the subject he talks of the animal machine being governed by three principal regulators – respiration, transpiration and digestion – he is clearly talking in descriptive terms. All he does is to state that these three activities are involved in most of the manifestations of life.

[1] Lavoisier was, however, aware that some sort of 'compensatory mechanisms' – which were in fact later discovered by Claude Bernard – existed in animals, though inevitably he was unaware of their exact nature. *See* 55, reprint p. 47.

'The effects of these different factors vary following an infinity of circumstances, even within fairly wide limits; and this is so, because in a variety of ways these regulators are self compensating, naturally attaining that equilibrium and regularity which constitutes good health.' (a)

Crawford, though holding similar views to Lavoisier, seems to commit himself a little further. After an experiment on a frog he speaks of its 'vital powers' as 'active in the generation of cold'.(b) We might of course interpret this purely as a description, and avoid the inference that Crawford regarded 'vital powers' as an additional explanatory agency, if it were not for the fact that on the very next page he describes an experiment 'to determine whether heat was lost by evaporation, or by the energy of the vital principle'. (He comes to the conclusion that the power of producing cold does not depend solely on evaporation.) Yet if Crawford did believe in a 'power' or 'principle' governing and explaining all vital activities he never specifically says so. All he does is carefully to point out the ways in which a living organism is able to regulate its own heat supply and other properties; and in this connexion, he notices an important point, which was not fully appreciated again until the work of Pflüger at the beginning of the twentieth century – namely, that the rate at which heat is produced by an animal is not uniform, but proportional to its needs.(c) The implication of this observation, though it was left to Claude Bernard to state it explicitly, is that 'combustion' in an inorganic body and in an animal are *not* completely analogous processes. Crawford was a mechanist in one sense, that he unhesitatingly used physical and chemical methods when studying living material; yet he would never have gone so far (one suspects) as to say that a complete explanation of the activities of living organisms could be given in terms of physics and chemistry alone.

Plenty of people at this time, particularly among doctors, were ready to question outright the relevance of chemical theory to physiology; and some of them even argued that no explanation in chemical terms should ever be attempted.

(a) 55, reprint p. 47. (b) 23, p. 384. (c) 23, p. 393.

Even the chemist, Chaptal,[1] who was in most respects a faithful follower of Lavoisier, was not prepared to allow that the methods of physics and chemistry could be applied to the study of living material without qualification. In the third volume of his *Elements of Chemistry* he says:

> 'The abuses which, at the beginning of the present [eighteenth] century were made of the applications of chemistry to medicine, have caused the natural and intimate relations of this science with the art of healing to be mistaken . . .
>
> 'In order to direct with propriety the applications of chemistry to the human body, proper views must be adopted relating to the animal oeconomy itself, together with accurate notions of chemistry itself. The results of the laboratory must be regarded as subordinate to physiological observations. It is in consequence of a departure from these principles that the human body has been considered as a lifeless and passive substance . . .
>
> 'In the mineral kingdom everything is subject to the invariable laws of affinities. No internal principle modifies the action of natural agents . . .
>
> 'In the vegetable kingdom, the action of external agents is equally evident; but the internal organisation modifies their effects . . .
>
> 'In animals functions are much less dependent on external causes and nature has concealed the principal organs in the internal parts of their bodies as if to withdraw them from the influence of foreign powers. But the more the functions of the individual are connected with its organisation, the less is the empire of chemistry over them and it becomes us to be cautious in the application of this science to all phenomena which depends essentially on the principles of life.' (a)

Here Chaptal is pointing to the variability and internal control of animal life, the principles of which (as he sees) can never be exhaustively understood in chemical terms alone; and which on this account he regards as escaping the laws of chemistry. In due course the distinction Chaptal draws here – between the living world inside the organism and its dead environment outside – was to become as rigid and absolute

[1] Chaptal was Professor of Chemistry at Montpellier, where a famous vitalistic body of opinion developed at this time.

(a) 20, p. 279.

as the earlier distinction in cosmology between the sublunary and superlunary worlds. The processes which take place inside the organism were supposed to be essentially different from those which take place outside it. In the case of the environment, the laws of physics and chemistry held good in all their deterministic rigidity; inside, the vital processes were governed by quite different principles. Essentially plastic, variable and constructive, these vital processes (which were the proper study of physiology) were contrasted with the determined, constant and destructive processes characteristic of physics and chemistry. This distinction was held more or less sharply by most of the positive vitalists of the next half-century; even some of the greatest were unable to see their way clearly beyond it. Liebig, though extending to physiological processes the determinism of the outer world, kept 'vital forces' of a sort unknown in the physico-chemical world outside as an element in his explanations. Many others, though rejecting the distinction between physical and vital force, still could not bring themselves to accept a complete determinism in physiology. And the destruction of this distinction was so difficult that its final breakdown – at the hands of Claude Bernard – deserves to be called the 'Copernican Revolution' of physiology.

Chaptal does not realize that, when we have rejected chemistry and physics as capable of giving us an explanation, we may be left with only a description. Berzelius, whose views are in other respects similar, realizes that this is scarcely enough:

'We may consider the whole of the animal body as an instrument which, from the nourishment it receives, collects materials for continual chemical processes, and of which the chief object is its own support. But with all the knowledge we possess of the forms of the body considered as an instrument, and of the mixture and mutual bearings of the rudiments to one another, yet the cause of most of the phenomena within the Animal Body lies so deeply hidden from our view, that it will certainly never be found. We call this hidden cause vital power and like many others, who before us have in vain directed their deluded attention, we make use of a word to which we can fix no idea. When the text-books inform us that the vital power in one place produces from the blood the fibres of the muscle, in another the bone... we know after this explanation as little as we knew before.' (a)

No wonder there is a tinge of exasperation as he writes that so far as explanation goes, references to 'vital power' take us no farther. But Berzelius at any rate tries to find a location for this unknown force or hidden cause.

'This unknown cause of the phenomenon of life is principally lodged in . . . the nervous system, the very operation of which it constitutes. Nothing of which Chemistry has taught us hitherto has the smallest analogy to the operation of the nervous system or affords us the least hint towards a knowledge of its occult nature . . . And the chain of our experience must always end in something inconceivable; unfortunately this inconceivable something acts the principal part in Animal Chemistry, and so enters into every process – even the most minute, that the highest knowledge to which we can attain, is a knowledge of the *nature* of the productions, whilst we are for ever excluded from the possibility of explaining *how* they are produced (b).'

Though Berzelius's vitalism does not lead him to reject (as Bichat did) the physiological explanations so far provided by chemistry, he is especially emphatic that, with regard to the working of the brain, chemical explanation is inadequate and always will be.

'But still more astonishing are the operations of the brain . . . Is it not probable that human understanding, which is capable of so much cultivation . . . may one day explore itself and its nature? I am convinced that it will not.' (c)

Xavier Bichat, to whom we now turn, has often been trounced for his vitalistic views, and often with a quite unjustified degree of venom. His greatest contribution to biology lay in his insistence that the causes of vital phenomena must be looked for in the properties of the individual tissues, and he did much to stimulate the work on tissue structure and properties which led to the cell theory. But his emphasis on 'vital properties' led his critics to believe that he regarded these basically as explanatory agencies or entities. This, I shall

(a) 9, p. 4. (b) 9, p. 5. (c) 9, p. 8.

argue, was a mistake – he used the term descriptively, and was a vitalist only in a *negative* sense alone. Bichat may be criticized with more justice however, over his uncompromising rejection of physics and chemistry as a tool for physiological investigation; although even here I am not sure that such criticism is not misplaced if one considers him in the context of his own times. The chief reason for his views was his absolute preoccupation with the variability of living material, and the necessity for taking this into account when giving general explanations of the workings of the organism. This preoccupation did not prevent him and his colleagues from making *positive* contributions to the growth of a scientific physiology, though it may have prevented them from contributing to the progress of biophysics and biochemistry.

Bichat utterly rejected any vitalist explanation which ascribes a unique principle or constituent to the living organism.

> 'This principle, termed vital by Barthez, archaeus by Van Helmont, is an assumption as void of truth as to suppose one sole acting principle governing all the phenomena of physics.' (a)

He considered that for the physiologist the existence of vital properties such as sensibility and contractility was as primary a deduction from observation as the existence of electricity and gravitation was for the physicist. The physiologist's business was to study the manifestations of these properties, and to relate them to the behaviour of the living tissues which was their effect. This, indeed, was all the physiologist could in his opinion be called upon to do. Bichat 'feigns no hypotheses' about the 'hidden nature' of the vital properties. He uses them to explain the action of a muscle [say] only as a physicist might appeal to 'gravity' to explain the falling of an apple.

In view of his almost positivistic remarks, it is surprising that Bichat has been described in retrospect (by Claude Bernard's biographer, Faure) as 'clinging instinctively to animistic ideas' (b) and (by Bernard's successor at the

(a) 10, p. v. (b) 35, p. 143.

Sorbonne, Paul Bert) as 'giving asylum to . . . a capricious force . . . which made all acts performed by living beings a series of miracles' (a) – an assessment of his work which is still widely accepted. Claude Bernard himself was more just in his estimate:

'In place of the metaphysical conceptions which had reigned hitherto [a reference to Barthez], here [in Bichat] is a physiological conception which tries to explain vital phenomena by the properties of the material tissues or organs themselves . . . Bichat may have been mistaken in his theory of life; but he was not mistaken in his physiological method.' (b)

What Bernard does object to is Bichat's idea that it is the function of 'the totality of vital properties' to resist the action of physical and chemical forces and properties on the organism.

In just one crucial respect Bichat's vital properties were unlike all physical properties. He felt that there might be laws of vitality, certainly, but that these had to be different from the laws governing the behaviour of inanimate matter, because of the variability of their action. This view committed him to a method of investigation totally unlike that followed by Lavoisier. Over this methodological issue he never compromised:

'One calculates the return of a comet, the speed of a projectile; but to calculate with Borelli the strength of a muscle, with Keill the speed of blood, with Lavoisier the quantity of air entering the lung, is to build on shifting sand an edifice solid itself but which soon falls for lack of an assured base. This instability of the vital forces marks all vital phenomena with an irregularity which distinguishes them from physical phenomena remarkable for their uniformity. It is easy to see that the science of organized bodies should be treated in a manner quite different from those which have unorganized bodies for object.' (c)

Of course a determinist might draw different conclusions from the same point. Liebig implicitly made this same criticism of Lavoisier; he argued that attempts to generalize about the

(a) 6, p. xiv. (b) 8, p. 161–2.
(c) 11, p. 81: for a complete account of his methodology, *see also* 10, ch. 1.

nature of respiration from quantitative experiments on the oxygen entering the lungs, and so on, were a waste of time, because of the many factors causing this quantity to vary.

Elsewhere Bichat writes:

> 'The laws of natural philosophy are constant and invariable; they admit neither of diminution nor increase . . . on the contrary the vital properties are at every instant undergoing some change in degree and kind; they are scarcely ever the same . . .
>
> '. . . In their phenomenon nothing can be foreseen, foretold nor calculated; we judge only of them by their analogies, and these are in the vast proportion of instances extremely uncertain . . .
>
> '. . . to apply the science of natural philosophy to physiology would be to explain the phenomena of living bodies by the laws of an inert body. Here, . . . is a false principle.' (a)

The variability of living phenomena must reflect, he felt, an underlying variability in the cause of these phenomena. This view he reiterates when considering animal heat:

> 'I shall observe that I have not attempted to decide in what manner caloric is produced in the capillary system . . . nothing of this can be submitted to experiment . . . It has been tried of late to determine precisely what quantity of oxygen was absorbed, what was required to produce the water of respiration . . . This accuracy would be advantageous if it could be attained but not a single phenomenon of the animal oecomony can admit of it in the explanations they give rise to; chemists and those in pursuit of natural philosophy, accustomed to study the phenomena over which physical powers preside, have transferred their theoretical calculations to the laws of vitality but it is no longer the same thing . . . The mode of theorizing with regard to the organized bodies must be quite different from that of theories applied to natural philosophy . . . Every physiological explanation [by contrast] should present nothing more than outlines or approximations; it should be vague if I may be allowed the term.' (b)

Yet earlier in the same section he had implied that we should expect to find some uniformity in the vital phenomena of the animal body:

(a) 10, p. xx. (b) 10, p. 621.

'There are phenomena [he says first] invariably united to the immutable order originally established, and which it is impossible to explain, only it appears that this order depends on the primitive type impressed upon the vital powers, a type which continually produced phenomena nearly uniform; but as numberless causes will occasion them to vary so will pulsation, respiration, etc . . .' (a)

This passage, implying a certain uniformity in physiological processes, does not to my mind contradict Bichat's general emphasis on the variability of the vital powers. He repeatedly insists on all the infinite variations and self-regulatory capacities of the human body, and here he is saying much the same as had Lavoisier when he spoke of the 'regulators' which compensate for external changes on the body. Doctors like Bichat, who were studying living creatures closely, and whose daily clinical work brought them up against all these variations, naturally distinguished sharply between the varied phenomena of life and the rigid regularities in the behaviour of inanimate material. It is not surprising, therefore, that they should have believed that different methods were necessary for investigating these two dissimilar types of being. Moreover, since the contrast between the living and the dead state was constantly before their eyes, they could not help but emphasize the destructive effect of physical and chemical forces which in life (they thought) were held at bay by living forces. Bichat's definition is famous: 'life is the sum of all those forces which resist death'. (b) But to suppose – as Magendie did – that he was offering this as anything more than a description is to do him an injustice.

Unfortunately Bichat died young; otherwise one feels that he might eventually have anticipated the views of Claude Bernard. He might have realized that the 'internal environment' was as capable of a rigid determinism – the 'immutable order' of which he spoke – as anything in the realms of physics and chemistry. But before leaving Bichat we must see what he had to say about the production of animal heat. This was amazingly modern:

(a) 10, p. 620. (b) 11, p. 1.

'Every one is acquainted with the innumerable theories set forth by certain physicians (mechanicians) respecting the production of animal heat. Modern chemists, in proving the insufficiency of such theories, have substituted another, attended with no less difficulty. The lungs have been considered by these as the focus from whence heat proceeded; and the arteries as tubes that serve to convey it throughout the body. According to this, the production of this important phenomenon belongs exclusively to the capillary system of the lungs. I believe, on the contrary, that it takes place in the general capillary system . . .

'The blood receives from two principal sources the substances that repair the losses it has sustained; these are – 1st. Digestion; 2nd. Respiration . . . Now, these new substances incessantly bring with them in this fluid a supply of caloric; for, as all bodies are impregnated with this fluid, there can scarcely exist an addition of substance without an addition of this principle. In hematosis, then, the caloric is combined with the blood, but is not disengaged; it forms a part of the fluid; it is one of its elements.

'Thus loaded with combined caloric, the blood reaches the capillary system; it is there distributed throughout where it becomes changed. It is, in fact, in this system that it is converted into the nutritive substance – into those of the secretions, the exhalations, etc . . .' (a)

It is difficult to quarrel with his account. Again Bichat's initial emphasis is on variability. When studying animals, he says, one notices four things: their capacity to maintain a temperature always above that of their environment; the local variations of heat that occur throughout the body; the fact that sometimes there seems to be a connexion between respiration, circulation and animal heat which at other times does not exist, and the fact that the tone of the muscles and the state of the nervous system influence heat production. All these facts (he says) show the inadequacy of existing physico-chemical analogies. Heat, he argues, must arise in the capillary system of the general circulation when required: otherwise how does one explain the local variations of temperature within the body? This was something that neither Lavoisier nor even Crawford could explain. Moreover, quite apart from

(a) 10, p. 605.

preoccupation with tissue properties. Bichat appreciated the relevance of the digestive processes to the production of animal heat. The role of food as a source of animal heat had been completely ignored by those chemists who accepted the theories of Lavoisier and Crawford. In this respect, indeed, Bichat was ahead of his time: the exact relation between digestion and heat production in the body had to wait for the quantitative methods of the great organic chemist, Liebig – and Liebig too was a vitalist.

By contrast let us now look at the views of two vitalists who offered a unique constituent or 'principle' as an explanation of the organism's activities. Blumenbach, a brilliant comparative anatomist and embryologist, held vitalistic views chiefly as a result of his embryological studies. His work appeared towards the end of the fierce controversy over preformation and epigenesis. Blumenbach talks of vital powers and principles as active agents in the body's economy.

'Although vitality is one of those subjects which are more easily known than defined and usually indeed rendered obscure rather than illustrated by an attempt at definition, its effects are sufficiently manifest and ascribable to peculiar powers only. The epithet vital is given to these powers because on them so much depends the action of the body during life and of those parts which for a short time preserve their vitality, that they are not referable to any qualities merely physical, chemical or mechanical.

'The latter qualities are of great importance . . . By physical powers . . . the rays of light are refracted to the axis of the eye. By chemical affinity the changes of respiration are effected. But the perfect difference of these dead powers from those which we are about to examine is evident from the slightest comparison of an organized economy with any inorganic body in which these inanimate powers are equally strong . . .

'. . . Vital powers are most conspicuously manifested by their resistance and superiority to others . . . During life they strongly oppose the chemical affinities which induce putrefaction . . . they so far exceed the force of gravity, that, according to the celebrated problem of Borelli, a dead muscle would be broken assunder by the very same weight which, alive, it could easily raise.' (a)

(a) 12, p. 16.

Blumenbach brought his vital principle even into his theory of animal heat.

> 'Since the changes [variations of the body] are effected by the energies of the vital powers only, the great influence of these in supporting our temperature must easily be perceived . . . Many arguments render it probable, that the action of the minute vessels is dependent on the varied excitement or depression of the vital principle, and the conversion of oxygenized into carbonized blood again upon this . . .' (a)

Blumenbach supposed that, superintending these vital powers, there was a central controlling agent; a formative force, 'a determined form designed for certain ends' the 'nisus formativus'. This force, in his view, directed morphogenesis, conserved the form of the organism throughout nutrition, and restored it if it were injured. But it had a constant, not a variable effect; this we recognize, he says, when we observe the organism. He emphasized this 'constant effect'– one which is presumably as rigidly determined in its own way as the effects of physical and chemical forces, though in its 'nature' different from these. This attitude, remarks Driesch (b), is one which Claude Bernard, as a determinist, would have been unable to criticize quite as fiercely as he did some other vitalist theories. But one wonders. When Blumenbach uses the term 'vital powers', he uses it as an *explanation* of the unique activities of the organism; when he speaks of his 'nisus formativus', he uses this too as an explanation of development – though in actual fact he does little else than give a general *description* of the main facts of development and growth. Even the constant effect of which he spoke so much, tells us little more than that embryological processes follow the same patterns each time. The difficulty is, of course, that there must be an explanation of the facts of development; that it must be possible to find some terms in which we can offer embryological explanation. But the problem remains; what are these terms, or what can be these terms? One sympathizes with Blumenbach and Hunter in their desire to go farther than Bichat and

(a) 12, p. 98. (b) 29a, p. 60.

Berzelius. But one cannot help contrasting the clarity with which Hunter and Blumenbach describe the conditions of the body and report their various experiments, with the uncertain terms in which their general explanations are expressed.

As we saw earlier, Hunter avoided committing himself to any particular theory of animal heat, though he did make accurate and careful experiments on heat variations in the body, and on the maintenance of bodily temperature. He objected to fermentation theories and respiration theories of animal heat, on account of their inability to explain the variations that he had observed in his clinical work; and once again he reverted to a 'general principle' as an alternative explanation.

> 'It is most probable that it [the cause of animal heat] arises from, some principle; a principle so connected with life that it can and does act independently of circulation, sensation and volition and is that power which preserves and regulates the animal machine.' (a)

He expands on the idea of this 'principle', but he is never able to do so with any precision or in a way free of ambiguity, and for this he was later to be severely criticized:

> 'By the living principle I mean to express that principle which prevents matter from falling into dissolution, – resisting heat, cold, and putrefaction, . . . I have asserted that life simply is the principle of preservation, preserving it from putrefaction.
>
> 'Animal and vegetable substances differ from common matter, in having a power superadded totally different from any other known property of matter, out of which arise various new properties; it cannot arise out of any peculiar modification of matter, but appears to be something superadded . . . Mere composition of matter does not give life, for the dead body has all the composition it ever had . . . This principle exists in animal substances devoid of apparent organization and motion . . . Organization and life do not depend the least on each other. Organization may arise out of living parts, and produce action; but life can never arise out of, or depend on, organization . . . Organization and life are two different things.' (b)

(a) 47a, p. 136.
(b) 47, footnote to pp. 120 and 121; *see also* 46, ch. II.

Yet Hunter at any rate was aware of the central difficulties of 'explanatory vitalism'.

> 'A bar of iron without magnetism may be considered like animal matter without life; set it upright, and it acquires a new property of attraction and repulsion at its different ends. Now is this any substance added? Or is it a certain change which takes place in the arrangement of the particles of iron giving it this property?' (a)

We find, then, a wide divergence of views around 1800, and two quite distinct vitalistic attitudes, combined with an almost universal tendency to oversimplify the issues involved in physiology. The methodology typical of the work of Lavoisier had hardly begun to make its impression; the descriptive vitalists refused to consider any explanation that did not include an account of the variability of organism; while the obvious and observable differences between living and non-living material led the explanatory vitalists to postulate the existence in nature of vital agencies of a non-physico-chemical kind. The resulting confusion of conflicting opinion could not be cleared up for more than half a century. Once again, the development of ideas about animal heat provides a useful index of progress in physiological method.

(a) 46, p. 222.

4

Theories of Respiration and Animal Heat up to the 1830s

CRAWFORD's theory of animal heat had been very favourably received. It appeared to rest on careful experiments and to be deduced strictly from facts. Because it was both complete and ingenious its fundamental points were all carefully examined. During the early part of the nineteenth century a number of experiments were performed, the results of which, though not directly opposed to those of Crawford, at least demonstrated that his conclusions were not fully established. Three fundamental points were further investigated: whether, as he thought, carbonic acid has a smaller capacity for heat than oxygen, so that heat is released when oxygen is converted into carbonic acid; whether the capacity of arterial blood for heat is greater than that of the venous; and whether the temperature of the blood is the same in both sides of the heart and in the main vessels of the pulmonary circulation.

The first of these points involved the general theory of combustion quite as much as the theory of animal heat. Exactly how heat came to be released during combustion was at this stage an argument in the field of chemistry – it was fundamentally as mysterious as all chemical reactions. Only much later, with the development of the general physics of energy-transfer, was it at all possible to throw fresh light on this side of the biological problem. Given that heat is produced during combustion, whether it was slow, as in the body, or fast, as in the burning of charcoal, the only problem for physiology was to find out where this combustion took place and whether there was a temperature gradient away from the source of the heat. Thus, though Crawford's figures for the specific heats of oxygen and carbonic acid were questioned early on in the

1800s (a), the essential merit of his theory remained, that it overcame the crucial objections to Black's theory.

One line of experimental work which now became lively and important was the attempt to measure the internal temperature of the various parts of the body. Many attempts had already been made to do this. It was first thought that the internal organs were so far below the surface, and so removed from the effects of the atmosphere, that they must all be at the same temperature. Hunter's measurements had indicated a difference in the temperature of organs according to their distance from the heart, and similar results were obtained by Sir Anthony Carlisle,[1] who in 1805 found a difference of 6° Fahrenheit between the temperature of the spleen and the bladder in a horse. (b) Humphrey Davy's brother, Dr. John Davy, found a temperature difference of 1½° between venous and arterial blood, and a similar difference between the temperatures of the right and the left ventricles. His experiments were made on four-month-old lambs:

> 'In each instance the animal was killed by the division of the great vessels of the neck; an opening was made immediately into the thorax, and a very delicate thermometer was introduced in the ventricles of the heart by means of a small incision. The operation occupied so short a space of time that in three instances the right auricle had not ceased contracting.' (c)

But the evidence was conflicting. Davy's conclusions were in contrast to those of Coleman and Cooper. Coleman (d) found the external temperature of both sides of the heart to be the same, and though the blood in the right ventricle was between one and two degrees warmer than that in the left, the blood in the left ventricle cooled more slowly, indicating that it contained more heat. Davy attempted to account for the discrepancies of these results, which were similar to those

(a) 27, pp. 72 and 113.

[1] In this paper Carlisle makes the point that the terms warm-blooded and cold-blooded are only relative because, as Hunter's researches had shown, all animals produce some heat. At this time, too, Spallazani's work showed that the gaseous changes that took place in animals with lungs took place in other animals too.

(b) 19, p. 1. (c) 25, p. 596. (d) 19, p. 15.

obtained by other physiologists, by relating them to the manner in which the animals were killed. In his experiments the animals were either killed by a blow on the head or by cutting into the vessels of the neck. Coleman and Cooper asphyxiated their animals. Davy noticed that after death by asphyxia there was generally an accumulation of blood in the right ventricle, and when the ventricle had been distended with blood he was able to detect little difference between the temperatures of the two sides. (a)

Those difficulties are not surprising, considering the technical problems involved. Experiments of this kind almost inevitably produced widely divergent results, and were open to a variety of interpretations. It was not possible to measure the internal temperature of a living animal directly; thermometers were neither small enough to be placed in the tissues without interference nor accurate enough to measure small differences. Accordingly observations had to be made directly over the point of the body in question; it was some time before thermo-electrical effects were recognized as a method of measuring temperature, and techniques were developed which enabled physiologists to measure internal temperatures directly, with the least possible interference to the body.

Davy's experiments on the 'specific caloric' of arterial and venous blood, the first made since Crawford, also gave different results. He concentrated on measuring the relative capacities of two bodies for heat rather than attempting to estimate their separate specific calorics independently. As he pointed out, it was difficult to obtain exact and accurate values for the specific caloric of the two types of blood, while it was comparatively easy to obtain an idea of their relative capacities; and for the theoretical point under discusssion this was quite as significant. Davy estimated the relative heat capacities by comparing the rates of cooling of two different types of blood, in addition to using mixture methods such as Crawford had used:

'The blood used was from the jugular vein and carotid artery of a lamb, about four months old. It was received in bottles, and to

(a) 25, p. 597.

remove the fibrin . . . was immediately stirred with a wooden rod . . . The specific gravity of the venous blood without the fibrin was found to be 1050, and that of arterial 1047.

'A glass bottle equal to 2518 grains of water in capacity, and weighing 1332 grains, was filled respectively with water and venous and arterial blood of the temperature of the room, 62°, about four hours after the blood had been drawn, during which time the bottle had been closely corked. A delicate thermometer was placed in the middle of the liquid, by means of a perforated cork. The bottle was then plunged into water of the temperature 140° Fahrenheit, and when the mercury had risen to 120°, the bottle was quickly wiped and suspended in the middle of the room, and the progress of cooling was noticed every five minutes, till the thermometer had fallen to 80°. The following were the general results obtained:

Water	cooled	from	120°	to	80°	in	91	mins.
Arterial blood	,,	,,	,,	,,	,,	,,	89	,,
Venous blood	,,	,,	,,	,,	,,	,,	88	,,

Considering then the capacity of water for heat to be denoted by 1000, neglecting the effect of the glass bottle producing a difference of only about half a minute, and being the same in each instance, and dividing the times of cooling by the specific gravity, the relative capacities of venous and arterial blood without fibrin, appear to be as ·921 and ·934.

'In the following experiments the same kind of blood and the same quantity were used as in the preceding. The mixtures were made in a very thin glass receiver containing a delicate thermometer . . . the quotient or specific caloric for venous blood appears to be as ·812, and for arterial as ·814.' (a)

Davy concludes that his results are in direct opposition to Crawford's hypothesis but are more 'agreeable and even support the hypothesis of Dr. Black'. They are also consistent, he says, with Brodie's influential hypothesis (which we will consider shortly) that animal heat is dependent for its production on the energy of the nervous system, and arises from 'vital actions'.

Davy also offers arguments against Crawford's view:

'As we never perceive a difference of capacity of bodies without a difference of form or composition; and as very slight differences of the

(a) 25, p. 592.

former result only from great changes of the latter, it might be expected *a priori*, as no difference, excepting that of colour, has been detected between venous and arterial blood, that their specific caloric would be very similar. From analogy also, it might be expected, that the capacity of arterial blood for heat would be much less than that of water, as water appears to exceed almost every other fluid, and as the capacity appears to diminish as the inflammability of compounds increases. But the strongest arguments against this hypothesis are to be derived from the recent experiments of Mr. Brodie, and those of MM. Delaroche and Berard.

'Dr. Black's hypothesis appears to me far more satisfactory than Dr. Crawford's, and capable of explaining a much greater number of phenomena; but there are objections even to this hypothesis which must be removed . . . if I were questioned which view is preferable, I should make no hesitation in selecting Dr. Black's, which appears to me both the most simple and the most satisfactory.' (a)

But once again, Davy's work did nothing to meet the objections to Black's theory. Nor did it affect the central point of the chemical theory, which Lavoisier had undeniably demonstrated: that, since a similar quantity of heat is formed when carbonic acid is produced during either combustion or respiration, heat production in animals is essentially connected with breathing.

The search for the exact connexion between animal heat and respiration was the immediate result of the work of Lavoisier and Laplace, and it demanded of course a more detailed study of the respiratory process itself. Much work was done and a large number of experiments were made but, as so often with physiological questions, the difficulties were tremendous. As Edwards points out, physiologists in 1832 were still agreed on only two points, and those were the ones made by Lavoisier half a century earlier; that a portion of oxygen disappears during respiration; and that some carbonic acid is produced. Such questions as 'How much carbonic acid is produced, compared with the oxygen used?' 'Is any oxygen absorbed into the blood?' 'How and where is carbonic acid formed?' and 'What, if any, is the role of nitrogen in respira-

(a) 25, p. 603.

tion?' remained in a great measure unanswered right until the 1850s – indeed, the gradual clarification in men's understanding of the respiratory processes during these years is a story in itself. Here we can discuss in detail only one of these questions, though it is perhaps the most fundamental of all; that is the question, whether any oxygen at all was absorbed into the blood. This issue was important because, once it was admitted that oxygen was actually taken into the blood unchanged, then the combustion which produced animal heat had to be recognized as being, not an external one in the lungs, but an internal one within the very tissues of the body. If this were so, one could once again argue (as Brodie was to do), that the process was fully under the influence of 'internal' or 'vital' forces, whereas followers of both Black and Crawford had hitherto assumed that the *combustion* was an *external* one, even though the actual release of heat might take place only in the capillaries.

During the early years of the nineteenth century it was generally taken for granted that some portion of the inspired oxygen was absorbed by the blood. Later this belief was almost universally discarded, but by 1836 opinion was swinging back again in favour of the original view. This vacillation of opinion closely followed the progress of experiments designed to show the changes that take place in air through respiration. We can trace the changes of ideas in the view of one man – Bostock. To begin with, in 1808, he advocated the older view in an interesting controversy with Ellis. Later, as a result of the experiments of Allen and Pepys, he, in his own words, 'became a convert' to Dr. Ellis's doctrine and 'supported it in my lectures' (a); but by 1836 he was 'convinced that the researches of Dr. Edwards, taken in conjunction with the former facts and analogies that were adduced, will cause us to revert to the [former] conclusion.'

Bostock first published his theory of respiration – his *Essay on Respiration* in 1804. (b) The following account is taken from his 'Remarks on Mr. Ellis's Theory of Respiration' printed in 1808 in the *Edinburgh Medical Journal*. As will be seen, he retained

(a) 16, p. 362 footnote. (b) 14.

the essential points of Crawford's theory, so far as the transfer of heat to the limbs was concerned.

'I shall begin by briefly stating the theory of respiration I regard as the most probable . . . The blood arrives at the right side of the heart in a venalized state, loaded with a quantity of the oxide of carbon; as it passes through the action of the air contained in the bronchial cells: a portion of the oxygen is removed from the air, part of which forming an intimate union with the oxide of carbon is expelled in the form of carbonic acid gas, while the remainder is dissolved in the blood. It is here necessary to remark that it is not oxygenous gas, but oxygene,[1] which is supposed to be mixed with the blood. The caloric thus set at liberty is employed, part of it in maintaining the temperature of the lungs, which would otherwise be cooled by the admission of the external air; part of it in carrying off aqueous vapour, and another portion in converting the carbonic acid into carbonic gas,[1] but the greatest part of it is united, in the form of specific heat, to arterial blood, which, by becoming arterialized, has its capacity for heat increased. The arterial blood is poured into the left cavity of the heart, and propelled through the arteries, into the extreme parts of the body. The oxygene which was dissolved into the whole mass of the blood, during the circulation, gradually unites itself more intimately to a portion of carbon in it, which it converts into the oxide of carbon, and thus the blood acquires the venous state. By this change, its capacity for caloric is diminished; the specific heat which it obtains in the lungs is given out in the capillary vessels, to keep up the temperature of the body, and the blood returns to the right side of the heart completely venalized.' (a)

In his book *An Enquiry into the Changes induced in Atmospheric Air*, published in 1807, Ellis raised several objections to this theory. His fundamental objection related to the passage of the gas through the coats and membranes of the blood vessels and bronchial cells, and this was to be a point of contention until the 1830s:

'When, therefore, air is said to enter the blood from the cells of the lungs, it must in some way be conveyed through the coats and cells of these blood vessels. After what manner therefore is it able to effect a passage?

[1] Bostock retains Lavoisier's view that a gas is formed only when caloric is *combined with* the principle of the gas.

(a) 15, p. 160.

'Every anatomist will allow, that the surface of the cells of the lungs is duly furnished with absorbent vessels, of which not only the ordinary absorption of fluids carried on by this surface but the frequent removal of morbid matter from the bronchial cells supply abundant proof... Does the air which is supposed to pass out of the cells of the lungs into the blood vessels by a process of absorption take the route of these absorbent vessels? To this question we reply, in the language of Haller, that the fineness of those vessels, the mucus perpetually smearing the surface of the cells, the elastic nature of the air itself, and its repulsion by water, so that it neither penetrates moist paper, cloth nor skin – all demonstrate that no air by this route gets in the blood...

'If, then, no proof exists of the passage of air into the blood by the ordinary course of the absorbent vessels, the only other mode of effecting this purpose that has been hitherto suggested is the power of chemical affinity. What then are the chemical affinities subsisting between venal blood and atmospheric air?' (a)

Ellis mentions the existing opinion of chemists that no chemical reaction occurs unless there is direct contact between the substances involved. He quotes earlier work, especially Priestley's, to show that colour changes can be produced in blood out of the body by exposing it to atmospheric air or oxygen, and that this change in the blood is accompanied by one in the air, with the formation of carbonic acid. (This gas was also produced during Humphry Davy's experiments, when venous blood was placed in contact with nitrous oxide.) These observations lead him to his second and third points:

'Does the carbonic acid which is here [in the foregoing experiments] met with, proceed ready formed from the blood, or is it in part formed from the air? No one has yet proved that any aeriform fluid, much less that carbonic acid, exists naturally in the blood, and if this be true no such aerial acid can be expected to issue from it. The carbonic acid also is not formed by the blood when it is confined in nitrogen gas, neither does the colour of the blood in that case undergo any sensible change; but this acid is formed by the blood, either in oxygen gas, nitrous oxide or atmospheric air, all of which are deteriorated thereby; whence it follows, that without the presence of oxygen gas, the blood is unable to form carbonic acid, and that this acid is

(a) 32, p. 117–18.

therefore in part formed out of that gas . . . We infer, therefore, from these experiments, that atmospheric air is decomposed by being placed in contact with venal blood, its oxygenous portion being in part converted into carbonic acid, and a quantity of its nitrogen being, in consequence, left free.

'But, supposing the air to be thus decomposed by the blood it still remains a question whether it has been first attracted by that fluid, then decomposed, and afterwards in part expelled; or, whether the decomposition has been affected without such previous attraction and admixture of air. The only evidence of this supposed attraction seems to be the small diminution of bulk which the air in all cases suffers; but this cannot be considered as proof of the attraction of the air; for it is a necessary consequence of that conversion of oxygen gas into carbonic acid which has been shown to take place when these substances are brought into contact. Even granting to the blood this power of attracting air, or its oxygenous portion, it is not easy to conceive why it should so readily lose it and again give out this air in the form of carbonic acid.' (a)

In reply to the objection concerning the necessity for direct contact, Bostock quoted Priestley's experiments, in which the same effect was produced when the blood was separated from the air by a bladder or a thick 'stratum of serum', as when the two were in direct contact. He agreed that the *mechanism* of this process remained unknown, but the facts of the case were in his opinion quite conclusive:

'It is sufficient to state that oxygen and blood can act upon each other through a membrane which is very much thicker, and probably much denser than that which separates the blood in the rete mirabile of the lungs from the air or the bronchial cells.' (b)

Ellis's second objection was an attempt to show that the exhaled carbonic acid could not come from the venous blood, since its presence there had never been shown directly, and it was never produced unless the blood had some access to oxygen. Bostock answers that this objection does not seriously affect his own hypothesis, and in return offers Ellis a more powerful and significant argument:

'He has indeed omitted a weapon that may be considered as a more decisive argument against the existence of carbonic acid in the

(a) 32, pp. 122–33. (b) 15, p. 161.

blood than any which he has adduced; viz. that the soda which is found in the serum is in the pure or caustic state. I therefore agree with him in his conclusion, that the carbonic acid which is exhaled from the lungs is formed in consequence of the decomposition of the atmospheric air that is received into them.' (a)

As Bostock rightly sees, all his theory requires is that some compounds of oxygen and carbon should be formed in the capillaries and (as we now know) these are for the most part loose carbonates of sodium.

On the third point, however – where and how the decomposition of the air occurs – Bostock does not feel that Ellis has stated the dilemma correctly:

'It is not supposed that the air is absorbed by the blood, and afterwards in part expelled, but that a portion of oxygen is attracted, which, after entering into a variety of new combinations, is only discharged as a constituent of some of these new compounds. This appears to be a sufficient reply to the only, or at least the principal objection which Mr. Ellis offers to my hypothesis, that "granting to the blood this power of attracting air, or its oxygenous portion, it is not easy to conceive why it should so readily lose it, and again give out this air in the form of carbonic acid. The only part of the oxygen that can be supposed to undergo this rapid succession of changes is the quantity which is necessary to convert the oxide of carbon into cabonic acid"; . . . Mr. Ellis again recurs to the objection that was noted above that the air and blood are not in contact in the lungs, but that the contact is necessary to chemical affinity . . . I think it sufficient once more to refer to Dr. Priestley's experiments, in order to prove that this action really does take place . . .' (b)

According to Ellis's hypothesis all the inspired oxygen was used in the lungs to convert carbon already existing in the venous blood into carbonic acid. The colour of the blood after it had been arterialized was due purely to the abstraction of carbon, not to the addition of anything. As Bostock showed, this hypothesis left several things unexplained. How did the change to a venous condition occur? This could not be due to the *acquisition* of carbon alone: most of the carbon in the blood was known to enter the blood-system by way of the thoracic

(a) 15, p. 162. (b) 15, p. 162.

duct, having originally been derived from the digestion. Yet no colour-change took place at that point in the system. Neither could the venalization be due to the abstraction of oxygen, because by this theory none was absorbed. Moreover, if Ellis's theory was correct, what interpretation could be placed on Priestley's experiments, which did at least demonstrate that blood could be arterialized under circumstances when no excretion of carbon seemed possible? Most important for Bostock – and this was his Achilles' heel – experiments on the quantities of carbonic acid produced in respiration had, up to that date, indicated a surplus of oxygen, whose disappearance had to be accounted for.[1]

Bostock's own theory, on the other hand, failed to explain the *mechanism* whereby oxygen was able to pass into the blood vessels. This he was willing to admit. Existing knowledge of both absorption and chemical affinity was insufficient to account for it. Nor, even with the admission that the gases did penetrate the tissues of the lungs, could he explain why, when they were travelling in opposite directions, they did not impede one another. Still less could he explain why one portion of oxygen was absorbed by the blood while another was not. It was difficult to see how this division of labour came about. Finally, his theory supposed that gases could change from the 'elastic' to the 'fixed' state in a surprisingly short space of time – as indeed they can.

The difficulties were apparently resolved by the experiments of Allen and Pepys, which set physiologists off on a false track. These experiments were believed to be extremely accurate, and their results appeared to show that during respiration in both men and guinea pigs 'the quantity of carbonic gas emitted is almost equal, bulk for bulk, to the oxygen consumed'. (b)

These results gave Ellis fresh impetus, and in the second

[1] Though Ellis in his reply to Bostock in the same volume of the *Edinburgh Medical Journal* quotes the experiments of Thompson and the opinions of Dalton, whom he had met, and with whom he had discussed this point, that an equal bulk for bulk exchange of oxygen and carbonic acid gas occurs in the lungs, so that no air need be assumed to enter the blood. (a)

(a) 33, p. 325. (b) 2, p. 404.

edition of his book (1811) he placed great weight on them. (a) In view of the difficulty of imagining a mechanism by which oxygen could be absorbed into the blood, experiments tending to show a volume-for-volume correspondence between oxygen and carbonic acid made it simplest to assume that there was a direct conversion of one substance to the other in the lungs. The change in colour of the blood after its passage through them would thus be due solely to the abstraction of carbon compounds. This view, though it still left many points unexplained, was popular, and the experiments adduced in its favour were striking enough to cause Bostock to alter his views. For some twenty-five years, accordingly, he accepted the idea of direct conversion in the lungs. But by 1836 work had been published which convinced both him and many other physiologists that oxygen was indeed absorbed into the blood.

The crucial work at this point was that of Faust and Mitchell, and it directly met Ellis's original objections to the absorption theory. In 1828 Dutrochet had published an account of phenomena which he christened the 'endosmose' and 'exosmose' of fluids. He found that, when two fluids of different densities were separated by a membrane taken from living material, a two-way flow of fluid occurred. The current moving into the membrane he termed 'endosmose'; the one moving out, 'exosmose'. He also found that this process could be reversed; if the internal fluid were less dense than the external, this was expelled and the fluid of greater density absorbed. (b) This phenomenon was afterwards shown by Thomas Graham of Glasgow to occur in gases as well: carbonic acid separated from coal gas by a bladder passed into the bladder until it was fully extended.

Faust and Mitchell, in 1830, designed a similar series of experiments which were of direct relevance to the lung problem. Results similar to Graham's were obtained by Faust with nitrogen and hydrogen, common air and carbonic acid, hydrogen and carbonic acid, and oxygen and nitrogen. In

(a) 32. (b) 30.

the latter case there was a simultaneous current in both directions. They used membranes taken from the crop of a domestic fowl. Faust saw that gases escape very slowly through most membranes even in the aeriform state, and concluded:

'. . . much less then will they [gases] be able to escape when deprived of their elasticity by solution in the blood. The last reflexion will oblige us to admit one of two things; viz. either that carbon and nitrogen are thrown into the air cells in liquid form, and then become gaseous, or that they are thrown out by the influence of exosmose in a gaseous form, the former united with oxygen. But as the former of these opinions has not a single fact . . . to support it, it must, we think, be rejected, while the latter view is confirmed by many facts; among which one may be here stated, viz. that the change of blood and gas occur, when the blood is contained in a bladder, and is, therefore, not under that vital influence by which the carbon might be imagined to be secreted in liquid form and thus reach the air.' (a)

Their most striking experiment was the following:

'(6.30 p.m.) A beef's bladder was filled with blood drawn from the *same animal* two hours beforehand which had been carefully preserved from contact with the air, by being kept in a large bottle perfectly filled, and closed with a ground glass stopper. The external surface of the bladder being washed, and wiped with a dry cloth, the whole was placed in the bottom of a large glass jar, which at one side was prevented from contact with the bladder, by the interposition of a piece of wood, thus allowing a circulation of air from the bottom to the top of the jar. A small jar half filled with lime water was now placed upon the bladder, and the external air was completely excluded by closely covering the mouth of the large jar. On examination at the end of forty minutes, a pellicle of carbonate of lime had formed on the surface of the lime-water. The density of this pellicle continued to increase until 11 o'clock, when, being broken, by slight agitation, the greater portion precipitated, a little still floating. At 12 o'clock the pellicle, having again become complete, was again broken and precipitated by agitation and the process was left undisturbed till morning. At this time a considerable quantity of carbonate of lime having been formed, the process was stopped. The formation of carbonate of lime in the vessel exposed over the bladder of blood was beyond all comparison, more rapid than in another jar

(a) 36, p. 30.

of lime water, exposed to open air. Now if the great thickness and density of texture of a beef's bladder just taken from the animal, be compared with the extreme tenuity of the tissue of the pulmonary capillaries, if the free circulation of liquid blood in the lungs be compared with the motionless and solid coagulum in the bladder, finally, if the high temperature of the lungs be considered, *there will, we think, not be any doubt of the absorption of oxygen and the escape of carbonic acid.* Nor must we forget the immense extent of the respiratory surface, and the constant renewal of air in opposition to the small surface of the bladder . . . and the unrenewed and stagnant air of the jar. If then oxidation and decarbonization of the blood occur under the least propitious circumstances how much more rapidly will they proceed when every condition conspires to promote their action. As, too, these changes proceed in a dead membrane, *they are partially, if at all, dependent upon vitality, and result principally, or wholly, from the agency of those physical influences which are the subject of this paper.*' (a)

The second paper, 'On the Penetrativeness of Fluids' by Mitchell, (b) showed that gases and liquids alike could pass in both directions through membranes recently extracted from animals, and through those in the living animal. Mitchell also found that substances of the sort we now term colloidal solutions would not pass through the membranes, and that in their case only water would go through. He concluded:

'. . . that the liquid, if penetrative, permeates a given tissue at a rate dependent on the character of the tissue and the power of penetration. If on the opposite side there exists a substance or power capable of occupying or removing it as fast as, or faster, than the membrane delivers it, the actual rate of transmission will be as high as possible; but if not so capable the accumulation will be at a lesser rate, and will represent the degree of permeability of the inviting substance alone . . . The power of this process in liquids, like that of gases, is not yet measured . . . a power marvellously great, but insusceptible of demonstrative reference at present to any know cause . . .

'The most striking generality is that of the high power of penetrativeness of gases for *organic molecular tissue*, long know to be infiltrable by liquids, but until now, not generally known to admit of any permeation, by at least *insoluble* aeriform substances.

'Secondly, we are struck with an unexpected result, the great POWER of gases to infiltrate *each other* . . . Solutions may now be

(a) 36, p. 31. (b) 69, p. 36.

esteemed infiltrations by solids and liquids of the tissues of each other' requiring perhaps only a fitness in size, rather than a chemical or cohesive attraction, for we see it subverting even the greatest cohesive power, and holding no apparent relation with known chemical affinities.

'The experiments on the mutual action of gases and liquids show that although a gas may, when alone presented to a liquid for which it has no chemical affinity, penetrate its molecular cavities, yet, it will again leave it to join any gas whatever, which is brought into communication with the liquid. Thus carbonic acid . . . readily penetrates blood or water, but returns from either into the air or any other gaseous substance which contains no carbonic acid . . . It is in this way, probably, that oxygen disappears and an exactly equal quantity of carbonic acid replaces it in the bronchial cells. Oxygen penetrates slowly the membranous tissue, to infiltrate and brighten the blood; carbonic acid is immediately formed, and being a gas differing from the remainder of the air yet in the air cells, its tendency is to return, to penetrate that air, and thus escape through the trachea along with it. The oxygen enters, because there is enough oxygen behind to permit that, . . . The carbonic acid formed, makes its escape, because invited by the molecular tissue of the atmospheric air . . . The investigations of John Davy, and our careful repetition of his experiments, . . . leave no doubt of the entire absence of carbonic acid in the blood. It must therefore be produced in one or two modes, either by the penetration of oxygen into the blood, and its union there with carbon, or the exit of carbon from the blood to unite with oxygen in the air cells. Now, as carbon is one of the most fixed substances in nature, and has not been proved capable of such a transmission, we are . . . compelled to adopt the other theory which is in accordance with the laws of gaseous infiltration.' (a)

These experiments removed much of the doubt whether it was possible for oxygen to penetrate into the blood *in the lungs*. But it had still not been demonstrated that oxygen could be found in the arterial blood *throughout the course of the circulation*, and it had been shown that carbonic acid as such was *not* present in the serum. It was therefore perfectly reasonable for Faust and Mitchell to assume, as they did, that carbonic acid was formed immediately oxygen entered the blood, and that this process occurred in the pulmonary capillaries themselves.

(a) 69, pp. 51, 53, 54, 56, 57.

The change in colour of the blood could still be attributed to the loss of carbon compounds, and this seems to be what they believed. It is true that they say in one passage that oxygen 'brightens the blood', but earlier in the paper they had said:

'We are still inclined to consider the black colour of venous blood as due to the presence of an excess carbon, free or loosely combined. It will not redden without the removal of this carbon, or its more intimate union with the other constituents of the blood.' (a)

Faust and Mitchell's emphasis on the purely physical as distinct from vital nature of this 'penetrativeness' is very important, since it provides one of the earliest demonstrations of a process occurring in an exactly similar way in both living and non-living materials. As the process was governed by purely physical forces, this was of course further evidence that living processes did depend on physical influences, even if they were not totally controlled by them.

The results of these experiments, taken in conjunction with others by Edwards, (b) caused many physiologists, including Bostock, to revert to the view that oxygen was absorbed by the blood, and that the characteristic gaseous changes took place within the tissues of the organism. Edwards experimented on many kinds of animal, and found that the proportion of oxygen consumed in respiration varied in volume from about one-third that of the carbonic acid produced, down to almost zero: the range of variations depended on the species of animal, its age, and constitution, and there was a great variation in the breathing of the same animal at different times. Edwards certainly showed that in many cases there was a disappearance of oxygen to be accounted for, but more important, his work demonstrated that the respiratory process was exceedingly complicated. It was therefore quite wrong to say, as Allen and Pepys (c) had concluded, that a volume for volume relation always existed between the two gases, and that therefore no oxygen was taken into the blood.

The views of Faust and Mitchell on the source of animal

(a) 69, p. 27. (b) 31. (c) 2, p. 427.

heat[1] are clear only in a negative sense: they did not regard it as directly dependent on the respiratory processes.

'Our theory does not account for the production of animal heat, but it is presumed that no well-informed physiologist now seeks for it in the action of the lungs, or in the process of de-carbonization. The simple fact that cold-blooded animals breathe without *any* increase in temperature proves that mere breathing to *any* amount will not produce heat. Like all other animal functions, that productive of heat is dependent on a normal condition of the blood, and is thus *indirectly* governed by the act of respiration. As in cold-blooded animals, there is no apparatus for producing heat, respiration does not in any way influence their temperature. So in some of the cases quoted by John Hunter, where blue-boys maintained a temperature preternaturally great, the blood was very imperfectly decarbonized. In such cases the caloric function found some novel stimulant.' (a)

Certainly by 1830 sufficient work had been done by Spallanzani and others to show, both that cold-blooded animals do have a definite temperature, and that they respire with similar gaseous exchanges whether or not they have lungs. The possession of lungs was evidently not necessary – as had hitherto been supposed – either for respiration or for heat production. Mere breathing alone, as Faust and Mitchell pointed out, would therefore not account for the production of heat. At this time, of course, the vaso-motor regulation of heat-loss in warm-blooded animals had not been suspected. It was not suggested until much later that the essential difference between warm-blooded and cold-blooded animals might lie less in the mechanism of heat production than in the way in which they regulated their heat *loss*. (This was another of Claude Bernard's fundamental insights.) But why do Faust and Mitchell presume that no well-informed physiologists believed animal heat to be produced in the lungs or 'in the process of decarbonization'? One can only assume that they

[1] Heat was still generally regarded as some form of matter; this was probably the view of most physiologists though they often did not enter into a discussion as to its nature. But Lavoisier's term 'caloric' is almost always employed. He would undoubtedly have approved of the analogy which Mitchell is attempting to draw since he himself refers to the 'penetrative' power of caloric.

(a) 69, p. 57.

had reached this conclusion because of a famous series of experiments which, between 1810 and 1814, had apparently undermined the whole chemical theory of animal heat. It is to these experiments that we must now turn.

In 1810 the chemical theory was attacked by Sir Benjamin Brodie, President of the Royal Society. He took for the subject of his Croonian Lecture 'The Influence of the Brain on the Action of the Heart, and on the Generation of Animal Heat'. (a)

Brodie had noticed that, after pithing, the contractions of an animal's heart could be maintained for ten to fifteen minutes, even though respiration had ceased. He also noticed that circulation could be forcibly maintained even if the head of the animal was removed and the blood vessels ligatured. He concluded, in accordance with the views of Cruickshank and Bichat, that the brain was not necessary to the action of the heart, and that if the brain was destroyed the circulation ceased only because respiration ceased. He suspected that, if respiration could be kept up artificially, he could produce continued contractions of the heart. This he did by fitting a small pair of bellows into the trachea of the animal under observation, and simulating respiratory movements. The bellows were so constructed that the lungs could be inflated and then allowed to empty themselves. He repeated the inflation every few seconds, keeping the lungs fully inflated with atmospheric air each time. His first experiment was designed to determine whether, under these artificial conditions, the heat of the animal would be maintained, and it was made on a dog.

'The temperature of the room was 63° . . . By having previously secured the carotid and cerebral arteries, I was enabled to remove the head with little or no haemorrhage. The artificial respirations were made about 24 times in one minute. The heart acted with regularity and strength . . . At the end of two hours the pulse had fallen to 70 (being 84 after 35 minutes, 88 after 1½ hours) and at the end of two hours and a half to 35 in a minute, and artificial respiration was no longer continued.

(a) 17, p. 36.

'By means of a small thermometer with an exposed bulb, I measured the animal heat at different periods.

'At the end of an hour the thermometer in the rectum has fallen from 100° to 94°.

'At the end of two hours a small opening being made on the parietes of the thorax, and the ball of the thermometer placed in contact with the heart, the mercury fell to 86°, and half an hour afterwards in the same situation it fell to 78° . . .

'On examining the blood in different vessels, it was found of a florid red colour in the arteries, and of a dark colour in the veins, as under ordinary circumstances.' (a)

Brodie repeated the experiment on a rabbit with the same results, and concluded:

'It has been a very generally received opinion that the heat of warm-blooded animals is dependent on the chemical changes produced on the blood by the air in respiration. In the last two experiments the animals cooled very rapidly notwithstanding the blood appeared to undergo the usual changes in the lungs, and I was therefore induced to doubt whether the above mentioned opinion respecting the source of animal heat is correct. No positive conclusions however could be deduced from these experiments. If animal heat depended on the changes produced on the blood by the air in respiration, its being kept up to the natural standard or otherwise, must depend on the quantity of the air inspired, and on the quantity of blood passing through the lungs in a given space of time: in other words, in proportion to the fullness and frequency of the pulse and fullness and frequency of the inspirations.' (b)

Brodie therefore repeated his experiments on a small dog and rabbits, whose pulse was about 130 and 140 a minute, and whose respirations were about 30 to 35 a minute; and he tried to make the artificial respirations correspond as closely as possible to the observed natural ones. In all these experiments the animals cooled rapidly, even though respiration was continued.

Brodie next tried to determine whether any heat was being produced by his artificial respiration. He took two animals of the same size and colour and decapitated them both, keeping

(a) 17, p. 38. (b) 17, p. 40.

up artificial respiration on the one animal, and recording the rectal temperature of both after the same intervals of time.

'In this experiment, the thorax even in the dead animal cooled more rapidly than the abdomen. This is to be explained by the difference in the bulk of these two parts. The rabbit in which circulation was maintained by artificial respiration cooled more rapidly than the dead rabbit, but the difference was more perceptible in the thorax than in the abdomen. This is *what might be expected if the production of animal heat does not depend on respiration, since the cold air by which the lungs were inflated must necessarily have abstracted a certain quantity of heat,*[1] particularly as its influence was communicated to all parts of the body, in consequence of the continuance of respiration.

'It was suggested that some animal heat might have been generated, though so small in quantity as not to counteract the cooling powers of the air thrown into the lungs. It is difficult or impossible to ascertain with perfect accuracy what effect cold air thrown into the lungs would have on the temperature of an animal under the circumstances of the last experiments independently of any chemical action on the blood: since, if no chemical changes were produced, the circulation could not be maintained, and if the circulation ceased, the cooling properties of the air must be confined to the thorax and not communicated in an equal degree to the more distant parts.' (a)

Brodie accordingly repeated his experiments on two other rabbits. On this occasion artificial respiration was kept up in only one of them, and circulation was prevented in both by a ligature in the base of the heart. He found, on comparing the results of the two sets of experiments, that an animal whose circulation was maintained by artificial respiration on the whole cooled more rapidly than one whose lungs were inflated in the same manner after circulation had ceased. As he says,

'. . . this is what one would expect if no heat was produced by the chemical action of the blood.'

His general conclusion is, therefore, that

'When the influence of the brain is cut off . . . no heat is generated notwithstanding the functions of respiration, and the circulation of the blood continue to be performed, and the usual changes in the

[1] My italics. (a) 17, p. 45.

appearance of the blood are produced in the lungs. When the air respired is colder than the natural temperature of the animal, the effect of respiration is not to generate but to diminish animal heat.' (a)

(How Aristotle would have cheered!)

Two years later Brodie announced the results of further experiments. (b) In these he obtained similar results in animals whose brain functions had been impaired, not by decapitation, but by poisoning. As the sensibility of the animal was progressively impaired, so the power of generating heat dropped. If artificial respiration were maintained and the animal allowed to recover from the poison, it progressively recovered the capacity to produce heat. But it could maintain its temperature above that of the environment only when the nervous energy was fully restored. He confirmed that the usual chemical changes took place both in the lungs and in the blood during artificial respiration, taking for standard the results obtained by Allen and Pepys:

'. . . every cubic inch of carbonic acid requires a cubic inch of oxygen for its formation and secondly, when respiration is performed by a warm blooded animal in atmospheric air, the azote remains unaltered, and the carbonic acid exactly equals volume for volume the oxygen gas which disappears.' (c)

Once again, Brodie found, the effect of artificial respiration was to cool the poisoned animal more rapidly than a dead one.

Brodie now turns to consider the theory of animal heat.

'The facts now, as well as those formerly adduced, go far towards proving that the temperature of warm blooded animals is considerably under the influence of the nervous system; but what is the nature of the connexion between them? Whether the brain is directly or indirectly necessary to the production of heat? These are questions to which no answers can be given except such as are purely hypothetical.

'We have evidence that, when the brain ceases to exercise its functions, although those of the heart and lungs continue to be performed, the animal loses the power of generating heat. It would, however, be absurd to argue from this fact that the chemical changes of the blood in the lungs are in no way necessary to the production of

(a) 17, p. 48. (b) 18, p. 378. (c) 18, p. 379.

heat, since we know of no instance in which it continues to take place after respiration has ceased . . . Of the opinions of Black, Laplace, Lavoisier and Crawford, it is proper to speak with caution and respect. But . . . it does not appear to me that any of the theories . . . proposed afford a very satisfactory explanation of the source of animal heat. Where so many and such various chemical processes are going on, as in the living body, are we justified in selecting any one of them for the purpose of explaining the production of heat?

To the original theory of Dr. Black, there is the unanswerable objection, that the temperature of the lungs is not greater than that of the rest of the system. To this objection the ingenious and beautiful theory of Dr. Crawford is not open; but it is still founded on the same basis with that of Dr. Black . . . and hence it appears difficult to reconcile either of them with the results of the experiments which have been related.

'It may perhaps be urged that as in these experiments the secretions had nearly, if not entirely, ceased, it is probable that the other changes which take place in the capillary vessels had ceased also, and that although the action of the air on the blood might have been the same as under ordinary circumstances there might not have been the same alteration in the specific heat of this fluid as it flowed from the arteries into the veins. But, on this supposition, if the theory of Dr. Crawford be admitted as correct, there must have been a gradual but enormous accumulation of latent heat in the blood, which we cannot suppose to have taken place without its nature being altered. If the blood undergoes the usual change in the capillary system of the pulmonary it is possible that it must undergo the usual change since these changes are obviously dependent on and connected with each other . . . We may moreover remark that the most copious secretions in the whole body are those of the insensible perspiration from the skin, and of the watery vapour from the mouth and fauces and the effect of these must be to lower rather than to raise the animal temperature. Under other circumstances also the diminution of the secretions is not observed to be attended with a diminution of heat.' (a)

Brodie also throws doubt on the relevance of Crawford's estimates of the specific heat of blood *in vitro*, as applied to blood *in vivo*, because of the effects of coagulation; and he ends his paper by emphasizing that he is not advancing any

(a) 18, p. 391.

positive or controversial opinions, but only wishes to state some obstinate experimental facts.

Brodie himself may not have wished to theorize on this matter, but many people were prepared to do so for him. In certain places, accordingly, one finds references to 'Brodie's hypothesis', that animal heat is produced by the action of the nervous system. This misreading can even be found in Claude Bernard. (a) Brodie in fact offers no hypothesis at all, but merely demonstrates that the respiratory changes are not in themselves the sole source of animal heat; all we conclude is that animal heat is 'under the influence' of the nervous system.

The effect of Brodie's work, demonstrating as he did the inadequacy of existing chemical explanations, was to focus the attention of investigators back on to the living body. For many people, these experiments seemed to spell the complete overthrow of Crawford's theory. For others, the work was significant because they chose to interpret the experiments as demonstrating a 'vital cause' of animal heat. We have only to re-read the passage from Berzelius about the difficulty of explaining brain function in terms of chemistry to realize that for a very long time some physiologists would be able to appeal to nervous phenomena as 'vital functions' withdrawn from the domain of physics and chemistry. John Davy comments on Brodie's work as follows:

> 'The last hypothesis which I mentioned, that which refers animal heat to vital action, has many facts in its support, especially the results of Mr. Brodie's . . . experiments; and the results of my enquiry are not incompatible with it. It may be said that the viscera of the thorax and abdomen are of highest temperature, because these parts are, as it were, the laboratories of life; and that the heat of the arterial blood and of the parts best supplied with this fluid, is greatest, because they lie deepest and abound most in the principle of life or vital action. This explanation was suggested to me by my brother Sir H. Davy.' (b)

It is curious to reflect that John Davy, the doctor, was prepared to select Black's chemical theory of animal heat in preference to any other, while Humphry Davy, the chemist,

(a) 7, p. 21. (b) 25, p. 602.

preferred a vitalistic explanation, because there was at that time considerable discussion among chemists about the chemical origin of animal heat. (a) For instance, Thomas Garnett, in his lectures at the Royal Institution in 1801 (b) spoke of animal heat as being due to the change of oxygen from the gaseous to the fixed state, this fixation taking place in three ways: through absorption by the iron in the red blood corpuscles (a hypothesis for which he had already convincing chemical evidence but which, as we have seen, was by no means accepted by all physiologists at that time); by the formation of carbonic acid; and by the formation of water. He retained Lavoisier's theory,[1] according to which oxygen gas is a compound of the oxygen base and caloric: (c) the formation of heat in the lungs was due (in his opinion) not to the difference in specific heats of various sorts of blood, but to the decomposition of oxygen gas.

(a) 24, pp. 445–50 for Davy's views on respiration.

(b) Listed under reference 24 in bibliography. Garnett's lectures were published in 1808 in the form of a small booklet, but this booklet does *not* carry the name of the author or lecturer. Garnett was Professor of Natural Philosophy and Chemistry at the Royal Institution at that time.

[1] *See Elements of Chemistry*, A. Lavoisier, Ch. VII; '*On the Decomposition of Oxygen Gas, etc.*' (Translation in 1790 by Kerr.) (c)

(c) 53, part 1, p. 78.

5

The Methodological Situation in the 1830s: 'Vital Principles'

THE first thirty years of the nineteenth century saw such a body of work done in physiology that the whole aspect of the subject changed. The precedent set by Lavoisier and Crawford was widely followed, and a vast amount of physico-chemical experimentation was done on living organisms. With this rapid extension of physical and chemical methods into the biological field, and with the obvious success of these procedures, few people by 1830 any longer defended Bichat's thesis that exact quantitative results were not to be hoped for from any experiments on living creatures.[1] Yet the theoretical side of the subject did not show the same progress, and the methodological problem which had preoccupied physiologists around 1800 remained for long in dispute. On the one hand, the influence of physical and chemical forces on vital activity was generally admitted; but on the other there still seemed to be aspects of physiology which resisted physico-chemical analysis; and the existence of properties unique to living organisms seemed to suggest the action of laws or forces which were also unique. The problem was to find the relation between the two realms – physico-chemical and vital – to find terms which would both allow for the plasticity of organic processes, and at the same time relate them to the familiar regularities

[1] Though the prejudice against any physiological experiments took a long time dying. Olmsted recounts how, in 1856, Trousseau gave a résumé of Brown–Sequard's work on the effects of extirpation of the suprarenal bodies together with his own clinical examination of a patient suffering from 'bronzed skin'. (Addison's disease – related to pathological changes in the suprarenals.) After the report, a Dr. Bouillard remarked that he found the clinical report fascinating and interesting but in the 'philosophical prologue' to the experiments of Brown–Sequard, 'nothing more than what one might call amusing physiology'. (a)

(a) 72, pp. 95–6.

of the inorganic world. Theories which involved appeal to special vital agencies or principles were felt by many to be unsatisfactory; yet for all the critical examination to which such theories were subjected it was not yet possible to put forward any effective account of physiological processes which entirely dispensed with them.

The regulatory processes, which Claude Bernard was later to study, provided as always a good case in point. Francis Delaroche, for instance, clearly stated the central dilemma in a paper in 1812 'On the Cause of Refrigeration observed in Animals exposed to a high Degree of Heat'. (a)

> 'The animal economy presents us with phenomena which differing in their nature from those exhibited by inorganic bodies cannot be explained by the ordinary results of the laws of physics; while at the same time it produces others, which, being more or less similar to physical effects, are apparently derived from the same laws.'

He criticized the physiologists like Bichat who,

> '. . . struck with the errors committed by those, who have had a rage for ascribing everything to mechanical laws, will not admit any explanation of this kind in the animal economy. They are of the opinion, that the phenomena essentially connected with the exercise of life must depend on the laws that govern vitality; and not on physical laws, which have little apparent connection with the former, and very frequently seem in opposition to them.' (b)

It is clear, he insists, that physical causes are operative in the body,

> 'And if some of the phenomena of life appear to be contradictory to those laws, to which inanimate bodies are subject, must we thus infer that it is the same with all of them? This reasoning, erroneous in itself, would be contradictory to experience. Who, indeed, can overlook the influence of physical causes in several of the phenomena of the animal economy; such as for instance distinct vision, which depends essentially on the refracting power of the humours of the eye . . . It is true [he continues] that physical causes alone are not sufficient to produce these results, and vital causes powerfully concur in them; but the influence of the former is not the less evident. Generally

(a) 26, p. 361. (b) 26, p. 362.

speaking it may be said, that there is scarcely a phenomenon of the animal economy, which is not owing to both. Sometimes the influence of physical causes is predominant, at others that of the vital; and frequently it is difficult to determine with precision what belongs to one and what to the other.' (a)

His own particular object of study, refrigeration, was a nice example of a mixed phenomenon; the cooling of the animal being a result both of external, physical influences and internal, vital ones.

'The production of cold, manifested in animals exposed to a high degree of heat, is the result of the evaporation of the perspirable matters; which in consequence of the increased action of the exhalant system, is so much the more considerable, in proportion as the external heat is greater. It is therefore at the same time the result both of physical and vital causes.' (b)

He does not wish his references to 'vital causes' to be misunderstood, and adds a footnote:

'When I speak of vital causes and vital laws, I do not mean to assert, that they are actually different from general laws, that govern inanimate matter, and independent of them; they are, perhaps only modifications of them; but I am of the opinion, that, in the present state of science, we must admit them, if we would acquire tolerably accurate ideas of the mode, in which the different functions of the body are executed. We are yet far from the period, when many of the phenomena exhibited by these bodies may be referred to the laws of physics.' (c)

Delaroche still retains to a certain extent the earlier distinction, enunciated by Chaptal – that physical causes are derivable from the world external to the animal, vital causes from within – but for him the central problem is to determine the relative scope of the two kinds of activity.

'It is of no small consequence, however, to attain this object; and the researches capable of leading to it may be ranked among the most important in physiology. If we can ever hope to acquire precise notions of vital powers, and how they differ from physical, it must be

(a) 26, p. 362. (b) 26, p. 374. (c) 26, p. 362.

by observing what is peculiar to them in the vital functions, not by vaguely ascribing to them all the phenomena of organic bodies....'(a)

Delaroche was wise to qualify his use of the phrase 'vital cause', because at this time indiscriminate appeals to vital principles were being widely criticized. The phrase 'vital principle', as it was often used, seemed only to exclude the possibility of any co-existing explanation in terms of physics and chemistry. But the methodological dilemma indicated by Delaroche was acute, and the arguments that raged over this particular phrase form a significant part of physiological literature at this time. One finds the working scientists and doctors roughly divided into four groups: those who would never use the phrase at any price; those who would never part with it at any price; those who – more moderately – felt obliged to retain it, although they would have liked to dispense with it; and, lastly, those who had no great objection to the phrase, but saw the need (as Harvey had done with the term 'spirits') to define the precise sense in which it was to be admitted.

Many people were, like Magendie, uncompromising:

'One of the most deplorable of all the illusions into which modern physiologists have fallen is the belief that, by inventing the phrase "vital principle" or "vital force", they have done something comparable to discovering universal gravitation.' (b)

One year after he had taken his degree, Magendie celebrated his entrance into the medical world with a scorching paper in which he denounced the deplorable state of physiology. He had been considerably influenced by Laplace, and was deeply dissatisfied with the indeterminate nature of physiological investigation. The main target of his attack was the fashionable distinction between the biological sciences and the 'exact' sciences of physics and astronomy, which had received their stable foundation as a result of Newton's work. His insistence on extending determinism to biology was intended to cut the ground from under the feet of those who, for methodological reasons, refused to experiment with living

(a) 26, p. 362. (b) 64, p. 14.

material. Once one admitted, as he did, that 'even in the existing state of science . . . whenever the vital force animates a body of given organization it will produce given phenomena', there could be no methodological objection to experiment. Nevertheless, though he never would admit vital *principles*, like many of his contemporaries, Magendie never could entirely dispense with the idea of a vital *force*.

'Why invent on occasion of each phenomenon of living bodies, a particular and special vital force?[1] Might we not content ourselves with a single force which we should call *vital force* in a general manner, in admitting that it gives rise to different phenomena according to the structure of the organs and tissues which operate under its influence? But is not even this single vital function too much? Is there not here a simple hypothesis, since we cannot detect it? It would be of greater advantage if physiology only commenced at the instant when the phenomena of living bodies become appreciable to our senses.' (a)

Throughout his life, as a reaction against the vitalist schools, Magendie retained a strong dislike of hypotheses, to such an extent that his very experimental work was affected. As Olmsted (b) points out, Magendie was afraid of planning his experiments, in case a preconceived idea affected his observations, and as a corollary, he had no objection to haphazard experimentation. (It was left to his pupil Claude Bernard to redress the balance.) But despite his opposition to any form of vitalism, he, himself, had no alternative on occasion but to use vitalistic terminology. As Flourens pointed out, too, Magendie belonged to no philosophical school – he might have said with Pascal 'We do not think the whole of philosophy worth an hour's trouble'. (c) His *forte* was experiment: but he could not for ever avoid an attempt at general explanations; and when he had to make them he was forced to maintain an attitude midway between the extreme vitalist and mechanist positions of his time. Had he felt able to, he would certainly have explained everything by physical forces alone – but he could not. Yet he always tried to separate what was physical in the organism from what was vital.

[1] He is here specifically attacking Bichat's doctrine of vital properties.

(a) 5, p. 7; *see also* 37, p. 95. (b) 71, p. 233. (c) 37, p. 103.

'Begin always by analyzing the phenomena, by isolating what is physical from what is vital . . . (a) Far be it from me to exaggerate[1] the importance of physical explanations . . . Thus, why is it that under the influence of a moral emotion, more or less vivid, we see the face redden or grow pale? There is here something peculiar, something which does not belong to the domain of physics.' (c)

Because Magendie felt it necessary to retain this distinction he criticized any attempt to assess vital force quantitatively. Quantitative measurement belonged to the realm of physics, and could be applied only to those phenomena in the body which were physical in character. For him it was an error in methodology to attempt to analyse the vital force in terms of physical analogies, as Liebig tried to do around 1840 when he attempted to extend these analogies to the point of making quantitative measurements of the 'force'. (d)

Perhaps Lawrence felt that even Magendie was going too far,[2] and that the only attitude respectable scientically was to admit that one simply could not give a satisfactory explanation.

'It is justly observed by Cuvier that the idea of life is one of those general and obscure notions produced in us by observing a certain series of phenomena possessing mutual relations and succeeding each other in constant order . . . We know not the nature of the link that unites these phenomena, though we are sensible that a connection must exist; and this conviction is sufficient to induce us to give it a name, which the vulgar regard as a sign of a particular principle, . . .

'We do not profess to explain how living forces . . . exert their agency. But some are not content to stop at this point . . . they wish to display the very essence of vital properties and penetrate to their first

(a) 64a, p. 14; 37, p. 103.

[1] Flourens, in his memoir of Magendie, insists that Magendie did exaggerate the part played by physical forces in the body. (b)

(b) 37, p. 103. (c) 59, pp. 122, 165; 64b, t.l., p. 202.

[2] Yet even Lawrence (e) could not avoid Bichat's difficulty. He points out that as there are so many phenomena peculiar to organisms, physiology is inevitably a science quite different, methodologically, from physics, even though organisms are subject to certain physical laws. Yet to use physical terms to describe living phenomena is, he says, to 'perpetuate false notions; similarly if one described chemical phenomena in physiological terms.

(d) 59, p. 160–1.

causes; they suppose the structure of the body to contain an invisible matter or principle by which it is put into motion . . . In showing that the motion of the heavenly bodies follow the same law as the descent of a heavy body to earth does, Newton explained a fact. The opinion under review is not an explanation of that kind; unless you find, what I am not sensible of, that you understand muscular construction better by being told that an Archeus, or subtile and mobile matter sets the fibres at work. This pretended explanation, in short, is a reference . . . to something that we do not understand at all; to something that cannot be accepted as a deduction of science, but must be accepted as an object of faith . . . I only oppose such hypotheses when they are adduced with the array of philosophical deduction . . . because they involve suppositions without any ground in observation and experience, the only sources of information on the subject.' (a)

As Lawrence quite rightly points out, the term 'life' can do little more than describe: the introduction of a phrase like 'vital principle' to explain the phenomena observed begs the question and is unhelpful. Bostock pointed out the linguistic difficulties very clearly.

'There are two senses in which the term principle has been correctly applied in natural philosophy; first, when we wish to designate a material agent, which produces some specific effect, as according to the doctrine of Lavoisier, oxygen is said to be the acidifying principle . . . or secondly, we may correctly employ the term principle to signify the cause of a number of phenomena, which essentially resemble each other, and which may be all referred to one or more general laws, as the principle of gravitation or the principle of chemical attraction. We may then inquire how far the term principle can be properly applied to the cause of the phenomena of life.

'I feel little hesitation in saying, that it cannot be used with propriety in the first sense, to designate any material agent, notwithstanding the high authority of those physiologists who maintain the existence of a "materia vitae", and go so far as to describe its visible and tangible properties; or of those who identify the cause of the characteristic properties of life with electricity or any analogous agent. Nor shall we find the term principle more appropriate when employed in the second sense, to express the supposed cause of a series of phenomena, which may be all referred to one or more general laws; for, according to the explanation which has been given of it by those who

(a) 57, p. 167.

have expressed themselves in the most intelligible manner, the vital principle has been employed to express all those actions which could not be referred to any other general principle. Besides the laws of mechanics and chemistry we observe in the living body various phenomena which essentially differ from these, and which we must therefore ascribe to some other cause; but we find that these phenomena differ essentially among themselves, so that if we make this want of resemblance the bond of union, we proceed upon the fundamentally erroneous plan of generalizing specific differences, or associating phenomena, not because they resemble each other, but because they cannot be reduced under any other class. We may then conclude, that when it is asserted that the blood resists decomposition in consequence of the operation of the vital principle, if the phrase have any definite meaning, it is saying no more than that the blood is not decomposed because it is contained in the vessels of the living body, an assertion which no one will be disposed to deny, but which unfortunately does not throw any light upon the subject of our investigation.

'I conceive that the present state of our knowledge does not admit of our giving a satisfactory answer to this question, but as far as we are able to understand it, I think it is very evident, that it depends upon no single cause or principle, but upon the conjoined operation of many actions, which together constitute life.

'The regular supply of fresh materials, as furnished by the digestive organs, the removal of various secretions and excretions, and lastly, the abstraction by the lungs of the superfluous carbon and water, effects which depend on the united agency of both chemical, mechanical, and vital actions, are among the various causes which probably all contribute to the ultimate object.' (a)

Bostock's views are interesting for two reasons; first, he sketches in outline an account of living phenomena which Bernard was to present thirty years later in greater detail. The only way, he argues, in which one can give a satisfactory 'explanation' of vital processes, which embraces all the facts but does not lead to misconceptions, is by limiting one's statements to an account of the mechanism of the various *activities* that go on within the organism, the totality of which make up life. This is what Bernard did. Second, Bostock is insistent that the particular linguistic role we demand of

(a) 16, pp. 402–5.

words should be carefully defined. In a footnote on 'The Hypothesis of the Vital Principle', (a) he analyses at great length its use by the physiologists of his time. His analysis is searching and very much to the point. If, he says, all they mean when they write that 'the vital principle bestows on bodies certain properties' is that animate matter possesses essentially different properties from inanimate matter, then no one will deny this. (This is a descriptive vitalism of a form that Claude Bernard never objected to.) But if they imply that the 'vital principle' is something that can be added to, or taken away from bodies, then not only are they going on beyond the limits of correct induction but, what is equally heinous, they are employing a form of speech that is thoroughly obscure. In the same way, when his contemporaries spoke of the vital principle as producing certain effects in the body, the phrase, if it was to have any meaning, 'must be intended to explain the mode in which the effect is performed, whereas [he complains] they only tell us that the effect in question is a result of vitality, a proposition the truth of which no one can doubt, but which affords us no insight into the nature of the operation'. The doctrine of the vital principle should be recognized as simply descriptive.

Much of the preoccupation of English physiologists with the doctrine of a vital principle was due to the extensive influence of Hunter. We have seen how inexact he was in expressing his views on this matter and, as Bostock put it, 'even the most devoted admirers of Hunter admit, that this eminent physiologist was not fortunate in the explanation of his principles'. (b) Those successors who tried in turn to explain Hunter's explanations often found themselves talking in an equally obscure manner. It was a fairly general opinion that Hunter could not have been as obscure as they often were; and one is inclined to agree with Bostock when he feels that Abernethy, for example, further confused matters by unconsciously attributing some of his own ideas to Hunter. Abernethy, whose notes on Hunter's lectures missed no chance of referring to the vital principle, introduced some variations

(a) 16, p. 403. (b) 16, p. 405 footnote.

on this particular theme that remind us of the earlier versatility of the idea of 'spirits':

'Irritability is the effect of some subtile, mobile, invisible substance superadded to the evident structure of muscles, or other form of vegetable and animal matter, as magnetism is to iron, and as electricity is to various substances with which it may be connected.' (a)

He believed that the nervous system was the home of this 'subtile, mobile substance', and finally concluded, with an eye on the fashionable phenomenon of galvanism, that the vital principle was something like electricity – even perhaps identical with it. (b)

Yet Bostock's critique did not completely rout the advocates of vital principles, any more than the earlier attacks by Philip (c) and Pritchard (d) had done, and the issue remained a live one, which every writer on physiology felt bound to touch on. (Bostock himself discusses some thirty books and papers on the subject, (e) and as late as 1867 we can still find Joseph Henry defending the idea in a note on vitality.) (f) A year after the appearance of the third edition of Bostock's *Elementary System of Physiology* (1836), James Palmer took up the debate again in notes to his edition of Hunter's *Works*. His allusion to Bostock's phrase 'the conjoined operation of many actions, which together constitute life' makes his purpose clear.

Palmer concedes that Hunter's use of the idea of a vital principle involves one in difficulties of interpretation; but some such idea, he is convinced, is quite indispensable to physiology. He rejects Magendie's attack on the phrase, applauding rather

'. . . the remark of Prout, "that it is absolutely necessary to assume the existence of some agency different from and superior to that which operates among inorganic matter" (Bridgewater Treatise, p. 443) whether this agency be called *life*, the *principle of life*, or merely *organic agencies*.' (g)

Surely, he remarks, if the only objection to the phrase lies in the fact that we are ignorant of the ultimate nature of life,

(a) 1, p. 39. (b) 1, p. 88. (c) 76. (d) 78.
(e) 16, p. 403d footnote. (f) 44, p. 387. (g) 74, p. 125.

equally good grounds exist for banishing the term principle from our vocabulary altogether. Actually he insists, the integrated functioning of the animal body clearly shows the insufficiency of explanation in physical and chemical ideas alone:

'The animal body may be regarded as an intricate piece of machinery, arranged in perfect accordancy with the pre-existing properties of matter, so as to admit of the agency of those properties for the accomplishment of its own ends . . . Physiology, in short, is a complex science, presenting phenomena which proceed from the combined operation of chemical, physical, and vital agencies, harmoniously intermixed, but in which the vital principle holds a sort of supremacy, directing and modifying all the subordinate agencies to its own definite ends, besides producing effects which are not referable to any other power.' (a)

This aspect of living things is inevitably ignored in any account which treats them as *aggregates* of physical and chemical systems.

'Now the objection to all these definitions is, that they deal merely with the effects arising from the presence of life, or of the vital principle, and do not touch the question at issue, viz. of the nature of the principle itself. Let us apply these sort of definitions to some analogous subject of inquiry, and we shall more readily understand this. Suppose, for instance, that a person wanted to ascertain the cause of motion in any complicated piece of machinery, as a watch or a steam-engine, and he were to be informed that "it was the combination of motions which resisted rest", or that it was a "collection of phenomena which succeeded one another for a limited time", or "that it was the result of the united actions and reactions of all the parts", etc., he would at once perceive that his informant either did not understand his question, or could not answer it.' (b)

Even the introduction of the term 'organization' does not really help matters. He felt, like Hunter before him, that no conceivable definition of life in structural terms alone could possibly do justice to the essentially dynamic and integrative features of vital activity:

(a) 74, pp. 124–5 footnote. (b) 74, p. 126 footnote.

'If by the term life we are to understand the complex assemblage of phenomena exhibited in a perfect animal, let it be so; in the same way as by the term movement we may understand the complicated actions and reactions in a piece of machinery: but let us not mistake such a definition for expositions of the original cause of the phenomena of life, or of the motions of machinery. Such a cause does exist, however, in the one case as in the other, and that quite independently of the particular form or structure of the animal or the machine, and perhaps independently of form or structure of any kind. Such definitions of life, therefore, as the preceding are plainly liable to the objection of making life dependent on organization and action, which appears to me to be nothing less than a complete reversal of the actual relations of cause and effect: for in regard to the cause, we cannot say that there is invariable antecedence or connexion; while in regard to the effect, it is not found invariably to follow the existence of the cause. We recognize its presence posteriorly to all action, as when an animal is frozen or asphyxiated. On the other hand, life is sometimes suddenly and completely extinguished, without any apparent change having taken place in the organization of the body.' (a)

It was agreed, then, that existing explanations were largely insufficient, but what else was one to admit? One could of course bet all one's money on the physico-chemical horse, but to do so at this stage would certainly be a gamble. Even when in 1865 Claude Bernard gave his physico-chemical account of the regulatory processes within organisms, it was still necessary for him to introduce a novel and uniquely *biological* concept before he could complete the picture. But in the 1830s there was still a gap in all existing explanations; this gap was called either ignorance or vital powers, principles or forces; and these in turn were even compared to 'the unknown quantities, x, y, z, as used in algebra.' (b)

But the attempt to prove that explanations in such terms must be respectable by comparing them with algebraic variables, rested on a completely fallacious analogy. 'Suppose "x" denotes the number of sheep,' says the mathematics master – and the not-so-stupid pupil replies, 'Please, sir, what if "x" isn't the number of sheep?' Using letters of the

(a) 74, p. 126 footnote.
(b) 16, p. 403 footnote; 74, p. 125 footnote.

alphabet for quantities that we are about to calculate is one thing; setting up scientific hypotheses is another. If we already know that there *is* a quantity to be determined (whether this is a number, or the strength of a force, or the amount of a substance) we may be able to compute its value algebraically; but to use these same letters to refer to things whose very existence is in doubt is quite a different matter. Bostock might well have answered, 'and suppose your vital principle isn't "x" – or anything else?'

If any effective use was to be made of the idea of a vital force, agency, or principle, the first necessity was to demonstrate that it *could* be the subject of definite and determinate observations. The attempt to do this systematically may – as we can perhaps now see – have been bound to fail. But the attempt had to be made: the usefulness or uselessness of the idea was something which, in the long run, had to be put to the test rather than simply argued about. Here lies the significance of the contrasting theories, and attitude of Justus Liebig, the distinguished German chemist who made a brilliant and systematic attempt to quantify the idea of vital force, and Claude Bernard, who showed at last how physiology could give adequate explanations of its subject-matter without using the term.

6

The Physiological Theories of Liebig and Bernard

BOOKS dealing with the solution of physiological problems could be of infinite size. There never seems to be a point comparable with (for example) the situation in physics after Newton's gravitational synthesis, when one could say, 'Well, from now on everything is obviously going to be different.' The study of physiology is characteristic not only for the speed with which the seeming solution of one problem throws up a chain of others, but also for the way in which all physiological processes in the body impinge on and affect each other; so that there is a constant wealth of variation, and no problem ever seems to reach a *final* synthesis. Nevertheless we can, from the perspective of our times, see stages where some sort of a synthesis was achieved; when results and ideas and methods were pulled together into a coherent framework. And in this chapter and the next we must trace how both the ideas of animal heat and the fundamental methodology of modern physiology were rounded off, and in some measure, completed, by the work of Liebig and Claude Bernard – one a vitalist, and the other a mechanist. To do this one cannot deal with the sequence in strict chronological order; the threads that these men drew together were taken from various fields which impinge ultimately on physiology – anatomical biology, organic chemistry and physics – as well as from within the subject itself.

One can conveniently begin with the work of the German chemist Liebig because, though there was little contact between the German chemist and the French physiologist (indeed, Bernard rarely mentions him), Liebig's chemical analyses were fundamental to most of Bernard's work, and antedated it. At the meeting of the British Assocation in 1840,

Liebig[1] presented the first part of a report on organic chemistry and its relation to agriculture and physiology. Two years later he published the second part of the report, which dealt more particularly with animal physiology, as the book *Animal Chemistry*. (a)

The appendix is perhaps the most important part of the work, because in it Liebig gives the analytical results which relate to the chemical processes of digestion and respiration. As John Webster remarked in his preface to the American edition of the book (1843), the whole force of Liebig's reasoning depended on the vast accumulation of these analytical results. Liebig had devised experimental methods by which he could determine the exact percentage composition of organic substances, and through his work, as Read puts it, 'organic chemistry attained the status of an exact science capable of mathematical treatment'. (b) He applied these methods to a study and analysis of foods, for example, the nature and quantities of fats, carbohydrates and so on, consumed by man and other animals; and used these chemical analyses as a basis for his physiological theories. Moreover he extended his work to the analysis of the body fluids and excretions, and to the chemical composition of the tissues, inferring the physiological role of the various organs from the chemical properties of the elements which made up their substance.

We have seen that one of the most difficult problems in physiology – and one that was still outstanding – was the question whether or not oxygen was absorbed by the blood. The experiments of Faust and Mitchell (c) were in no way conclusive on this point. They showed only that it was perfectly *possible* for gases to penetrate living material; whether they actually did so in the organism was another matter. As we shall see later on in this chapter, one line of attack on this problem concentrated on trying to extract the gases from the blood and measure their volume; but this was

[1] In 1839, at the age of twenty-six, Claude Bernard was admitted as an interne at the Hotel-Dieu. Liebig at this time was thirty-six.

(a) 59. (b) 79, p. 170. (c) 36 and 69.

a course which Liebig did not consider particularly fruitful. There were technical reasons why the amount of gas extracted would not be a reliable measure of the actual content in the blood:

'Now we observe that the globules of arterial blood retain their colour in the larger vessels, and lose it only during their passage through the capillaries. All those constituents of venous blood, which are capable of combining with oxygen, take up a corresponding quantity of it in the lungs. Experiments made with arterial serum have shown that when in contact with oxygen it does not diminish the volume of that gas. Venous blood, in contact with oxygen, is reddened, while oxygen is absorbed; and a corresponding quantity of carbonic acid is formed.

'It is evident that the change of colour in the venous globules depends on the combination of some one of their elements with oxygen; and that this absorption of oxygen is attended with the separation of a certain quantity of carbonic acid gas.

'This carbonic acid is not separated from the serum; for the serum does not possess the property, when in contact with oxygen, of giving off carbonic acid. On the contrary, when separated from the globules, it absorbs from half its volume to an equal volume of carbonic acid, and, at ordinary temperatures, is not saturated with that gas . . .

'. . . we find, in point of fact, that the living blood is never in any state saturated with carbonic acid; . . . it is capable of taking up an additional quantity, without any apparent disturbance of the function of the globules. Thus for example after drinking effervescing wines, beer, or mineral waters, more carbonic acid must necessarily be expired than at other times. In all cases, where the oxygen of the arterial globules has been partly expended otherwise than in the formation of carbonic acid, the amount of this latter gas expired will correspond exactly with that which has been formed; less, however, will be given out after the use of fat and of still wines, than after champagne . . .

'. . . All the compounds present in venous blood, which have an attraction for oxygen, are converted, in the lungs, like the globules, into more highly oxidized compounds; a certain amount of carbonic acid is formed, of which a part always remains dissolved in the serum of the blood.

'The quantity of carbonic acid dissolved, or of that combined with soda, must be equal in venous and arterial blood, since both have the same temperature; but arterial blood, when drawn, must, after a short time, contain a larger quantity of carbonic acid than venous blood, because the oxygen of the globules is expended in producing that compound.' (a)

Liebig relied rather on the chemical properties of elements and compounds in the blood to provide the clue to the processes going on during the course of circulation. He had found a compound of iron in the red blood corpuscles, and could find no trace of iron compounds elsewhere in the body tissues. He knew too, of course, that iron compounds possess characteristic properties of oxidation and reduction. He showed that the compounds in the red corpuscles could be decomposed by some gases, such as carbonic acid and sulphuretted hydrogen, whereas they would combine with others (oxygen and nitrous oxide) displacing the compounds formed by the former gases. Moreover, though these gases were absorbed by florid blood they were not absorbed by the serum of that blood when the corpuscles were removed. He therefore inferred that the oxygen taken into the lungs was, in fact, absorbed by the blood, displacing in the process carbonic acid compounds rather than pure carbonic acid as had formerly been supposed.

'During the passage of the venous blood through the lungs, the globules change their colour; and with this change of colour, oxygen is absorbed from the atmosphere. Further, for every volume of oxygen absorbed, an equal volume of carbonic acid is, in most cases, given out.

'The red globules contain a *compound of iron*; and no other constituent of the body contains iron.

'Whatever change the other constituents of the blood undergo in the lungs, this much is certain, that the globules of venous blood experience a change of colour, and that this change depends on the action of oxygen . . . (b)

. . . . .

'According to the views now developed, the globules of arterial blood, in their passage through the capillaries, yield oxygen to certain constituents of the body. A small proportion of this oxygen serves to produce the change of matter, and determines the separation

(a) 59, p. 258. (b) 59, p. 253.

of living parts and their conversion into lifeless compounds, as well as the formation of the secretions and excretions. The greater part, however, of the oxygen is employed in converting into oxidized compounds the newly formed substances, which no longer form part of the living tissues.

'In their return towards the heart, the globules which have lost their oxygen combine with carbonic acid, producing venous blood; and, when they reach the lungs, an exchange takes place between this carbonic acid and the oxygen of the atmosphere.

'The organic compound of iron, which exists in venous blood, recovers in the lungs the oxygen it has lost, and in consequence of this absorption of oxygen, the carbonic acid in combination when it is separated.

'Hence, in the animal organism, two processes of oxidation are going on; one in the lungs, the other in the capillaries. By means of the former, in spite of the degree of cooling, and of the increased evaporation which takes place there, the constant temperature of the lungs is kept up; while the heat of the rest of the body is supplied by the latter.' (a)

Now, believing that oxygen actually entered into the living system, Liebig realized that one should be able to give an exact chemical account of its subsequent disappearance. In doing this he answered a question which was absolutely fundamental to the solution of the problem of animal heat, but one which hitherto had not received much attention: namely, what was the origin of the carbonaceous matter which was subsequently burnt.

'At every moment of his life man is taking oxygen into his system. by means of the organs of respiration.

'The observations of physiologists have shown that the body of an adult man, supplied with sufficient food, has neither increased nor diminished in weight at the end of twenty-four hours; yet the quantity of oxygen taken into the system during this period is very considerable.

'According to the experiments of Lavoisier, an adult man takes into his system, from the atmosphere, in one year, 746 lbs., according to Menzies, 837 lbs. of oxygen; yet we find his weight, at the beginning and end of the year, either quite the same, or differing, one way or the other, by at most a few pounds.

(a) 59, p. 261.

'What, it may be asked, has become of the enormous weight of oxygen thus introduced, in the course of a year into the human system?

'This question may be answered satisfactorily: no part of this oxygen remains in the system; but it is given out again in the form of a compound of carbon or of hydrogen.

'The carbon and hydrogen of certain parts of the body have entered into combination with the oxygen introduced through the lungs and through the skin, and have been given out in the forms of carbonic acid gas and the vapour of water.

'At every moment, with every expiration, certain quantities of its elements separate from the animal organism, after having entered into combination, within the body, with the oxygen of the atmosphere.

'If we assume, with Lavoisier and Seguin, in order to obtain a foundation for our calculation, that an adult man receives into his system daily 32½ oz. (46,037 cubic inches = 15,661 grains, French weight) of oxygen, and that the weight of the whole mass of his blood, of which 80% is water, is 24 lbs.; then it appears, from the known composition of the blood, that, in order to convert the whole of its carbon and hydrogen into carbonic acid and water, 64,103 grains of oxygen are required. This quantity will be taken into the system of an adult in four days, five hours.

'Whether this oxygen enters into combination with the elements of the blood, or with other parts of the body containing carbon and hydrogen, in either case the conclusion is inevitable, that the body of a man, who daily takes into the system 32½ oz. of oxygen, must receive daily in the shape of nourishment, as much carbon and hydrogen as would suffice to supply 24 lbs. of blood with these elements; it being pre-supposed that the weight of the body remains unchanged, and that it retains its normal condition as to health.

'This supply is furnished in the food.' (a)

The realization that the carbonic acid gas breathed out must have come from the complex foodstuffs taken in was one of Liebig's greatest insights and, moreover, he was able to give an account of the chemical transformations which took place in the body during this conversion of food into the gas. His analyses of starch, fats, and so on eaten by animals showed him that these substances contained no ready formed carbonic acid, and that this gas, which was given out in the lungs, must

(a) 59, p. 13.

have been formed in the tissues. He was one of the first people to realize, and to state clearly, that animals do *not* have the power to synthesize their own foods, but that growth is possible and life maintained only by the breakdown into a soluble form of complex foodstuffs. This is followed by the oxidation of both the soluble compounds and (Liebig believed) the *tissues* of the body; the oxidation products are then chemically synthesized to form the tissues. The crucial tissue was, of course, the blood, which supplied to each organ everything material necessary for growth. Liebig believed that it supplied the materials in a chemically complete form; no synthesis took place in the tissues – only growth. The carbon dioxide and water excreted came from the oxidation of the tissues and soluble foodstuffs in the blood. Liebig knew that sugars could be oxidized to carbonic acid gas, water and alcohol by fermentation. Part of the sugar could be supposed to form alcohol, yielding oxygen as it did so, and this oxygen then oxidized completely the remaining third of the sugar.

[1](Two thirds) $(C_{29} H_{24} O_{24} + H_{12} O_{12}) - O_{24} = C_{24} H_{36} O_{12} =$ 6 equivalent alcohol.

(One third) $(C_{12} H_{12} O_{12} + O_{24} = 12\,CO_2 + 12\,HO$. (a)

Similar types of oxidation took place in the body. In all cases heat was produced whether oxidation was supposed to take place by the separation of carbonic acid gas and water from pre-existing compounds, as in fermentation, or by direct oxidation of sugar and other compounds.

For Liebig, as later for Claude Bernard, respiration and

[1]It will be noticed that Liebig gives a formula for sugar in which the number of hydrogen and oxygen atoms is the same. Dalton had assumed, in all his calculations of atomic weights, that the composition of water was HO – rather than H_2O. The ramifications of this mistake caused chemists a great deal of trouble right up to the 1850s, and rival tables of atomic weights remained in circulation until after 1858. In that year, Cannizzaro showed how, by reviving the neglected hypothesis of Avogadro (all simple gases at the same temperature and pressure contain the same density of molecules), one could introduce consistency into the conflicting tables – though at the cost of doubling the previous estimate of the atomic weight of oxygen and many of the metals. This change makes no material difference, of course, to Liebig's account of the decomposition of sugar, but only to his chemical formulae.

(a) 59, p. 88.

nutrition were complementary processes and the source of all the vital activity of the animal. But, unlike Bernard, he felt these processes to be influenced by a force unique to the animal.

'If the first condition of animal life be the assimilation of what is commonly called nourishment, the second is a continual absorption of oxygen from the atmosphere.

'Viewed as an object of scientific research, animal life exhibits itself in a series of phenomena, the connexion and recurrence of which are determined by the changes which the food and the oxygen absorbed from the atmosphere undergo in the organism under the influence of the vital force.

'All vital activity arises from the mutual action of the oxygen of the atmosphere and the elements of the food.' (a)

The process of respiration also creates the animal's heat – almost incidentally, Liebig implies. The heat is inevitably produced because, during respiration, a chemical reaction is taking place in the tissues of the body, and this reaction just happens to be an exothermic one.

'The mutual action between the elements of the food and the oxygen conveyed by the circulation of the blood to every part of the body is THE SOURCE OF ANIMAL HEAT.

'All living creatures, whose existence depends on the absorption of oxygen, possess within themselves a source of heat independent of surrounding objects . . .

'It is only in those parts of the body to which arterial blood, and with it the oxygen absorbed in respiration, is conveyed, that heat is produced. Hair, wool, or feathers do not possess an elevated temperature.

'This high temperature of the animal body, or, as it may be called, disengagement of heat, is uniformly and under all circumstances the result of the combination of substance with oxygen.

'In whatever way carbon may combine with oxygen, the act of combination cannot take place without the disengagement of heat. It is a matter of indifference whether the combination takes place rapidly or slowly, at a high or at a low temperature; the amount of heat liberated is a constant quantity.' (b)

(a) 59, p. 9. (b) 59, p. 17.

Liebig gave, then, an account of the chemical processes involved in the production of animal heat which related it to both the gases breathed and the food eaten. Since his time biochemists and physiologists have filled in a great deal of the chemical detail, but in general outline we still retain his views. But Liebig contributed something further; a new awareness of the problems of biochemistry. In the earlier days of physiology, theoretical conclusions had, of course, been drawn from experiments involving quantitative measurements; indeed, we have seen how Bostock switched to Ellis's theory of respiration largely because of Allen and Pepys's experiments on the quantities of oxygen consumed and carbonic acid gas produced by an animal. But one of the annoying peculiarities of these early experiments was their widely divergent results, which made it very difficult to arrive at any firm theoretical conclusions. Liebig showed that the problem had been oversimplified, and that, while quantitative technique must remain the basis of physiological procedures, greater care should be taken in interpretation of the results.

For example, he listed all the conditions that caused the number of respirations to vary – exercise, work, illness, starvation, heat, cold, pressure – and showed that, because of these, quantitative generalizations drawn from the standard sorts of experiment were of little relevance to the theoretical problem:

'All experiments hitherto made on the quantity of oxygen which an animal consumes in a given time, and also the conclusions deduced from them as to the origin of animal heat, are destitute of practical value in regard to this question, since we have seen that the quantity of oxygen consumed varies according to the temperature and density of the air, according to the degree of motion, labour, or exercise, to the amount and quality of the food, to the comparative warmth of the clothing, and also according to the time within which the food is taken . . .

'The attempts to ascertain the amount of heat evolved in an animal for a given consumption of oxygen have been equally unsatisfactory. Animals have been allowed to respire in close chambers surrounded with cold water; the increase of temperature in the water has been measured by the thermometer, and the quantity of oxygen consumed has been calculated from the analysis of the air before and after the experiment. In experiments thus conducted, it has been found that the animal lost about 1/10th more heat than corresponded to the oxygen consumed; and had the windpipe of the animal been tied, the strange result would have been obtained of a rise in temperature of the water without any consumption of oxygen. The animal was at the temperature of 98° or 99°, and the water, in the experiments of Despretz, was at 47·5°. Such experiments consequently prove, that when a great difference exists between the temperature of the animal body and that of the surrounding medium, and when no motion is allowed, more heat is given off than corresponds to the oxygen consumed. In equal times, with free and unimpeded motion, a much larger quantity of oxygen would be consumed without a perceptible increase in the amount of heat lost. The cause of these phenomena is obvious. They appear naturally both in man and animals at certain seasons of the year, and we say in such cases that we are freezing, or experience the sensation of cold. It is plain, that if we were to clothe a man in a metallic dress, and tie up his hands and feet, the loss of heat, for the same consumption of oxygen, would be far greater than if we were to wrap him up in fur and woollen cloth. Nay, in the latter case, we should see him begin to perspire, and warm water would exude, in drops, through the finest pores of his skin.

'If to these considerations we add, that decisive experiments are on record, in which animals were made to respire in an unnatural position, as for example, lying on the back, with the limbs tied so as to preclude motion, and that the temperature of their bodies was found to sink in a degree appreciable by the thermometer, we can hardly be at a loss what value we ought to attach to the conclusions drawn from such experiments as those above described.

'These experiments and the conclusions deduced from them, in short, are incapable of furnishing the smallest support to the opinion that there exists, in the animal body, any other unknown source of heat, besides the mutual chemical action between the elements of the food and the oxygen of the air.' (a)

Liebig was well aware that all animals produce heat and that, strictly speaking, they are all therefore 'warm-blooded'. He had also noticed that so-called warm-blooded animals are able to maintain their temperature above that of their environment, but can only do this by keeping up their supply of food.

(a) 59, p. 36.

From this observation he draws an analogy that is both striking and interesting

'The animal body is a heated mass, which bears the same relation to surrounding objects as any other heated mass. It receives heat when the surrounding objects are hotter, it loses heat when they are colder than itself . . .

'Now, in different climates the quantity of oxygen introduced into the system by respiration, as has already been shown, varies according to the temperature of the external air; the quantity of inspired oxygen increases with the loss of heat by external cooling, and the quantity of carbon or hydrogen necessary to combine with this oxygen must be increased in the same ratio.

'It is evident, that the supply of the heat lost by cooling is effected by the mutual action of the elements of the food and the inspired oxygen, which combine together. To make use of a familiar, but not on that account a less just illustration, the animal body acts, in this respect, as a furnace, which we supply with fuel. It signifies nothing what intermediate forms food may assume, what changes it may undergo in the body, the last change is uniformly the conversion of its carbon into carbonic acid, and of its hydrogen into water; the unassimilated nitrogen of the food, along with the unburned or unoxidized carbon, is expelled in the urine or in the solid excrements. In order to keep up in the furnace a constant temperature, we must vary the supply of fuel according to the external temperature, that is, according to the supply of oxygen.

'In the animal body the food is the fuel; with a proper supply of oxygen we obtain the heat given out during its oxidation or combustion.' (a)

The analogy with a furnace, and therefore with direct chemical combustion, is interesting because it was one which, later on, Mayer also was to draw while developing his views on the mechanical equivalent of heat. But, as we shall find Bernard remarking, its use lays a physiologist open to grave criticism.

Finally Liebig was able to explain Brodie's results, and disposed of the view that the origin of animal heat lay in the nervous system. In fairness to Brodie it must be noted that since 1812 anatomists had studied the innervation of the viscera and the effects of dividing the visceral nerves much

(a) 59, p. 18.

more completely, and Liebig could show that Brodie's results were a consequence of the influence of the digestive processes, which act indirectly on the process of respiration.

'The want of a just conception of force and effect, and of the connexion of natural phenomena, has led chemists to attribute a part of the heat generated in the animal body to the action of the nervous system. If this view excludes chemical action, or changes in the arrangement of the elementary particles, as a condition of nervous agency, it means nothing else than to derive the presence of motion, the manifestation of a form, from nothing. But no force, no power can come of nothing.

'No one will seriously deny the share which the nervous apparatus has in the respiratory process; for no change of condition can occur in the body without the nerves; they are essential to all vital motions. Under their influence, the viscera produce those compounds, which, while they protect the organism from the action of the oxygen of the atmosphere, give rise to animal heat; and when the nerves cease to perform their functions the whole process of the action of oxygen must assume another form. When the Pons Varolii is cut through in the dog, or when a stunning blow is inflicted on the back of the head, the animal continues to respire for some time, often more rapidly than in the normal state; the frequency of the pulse at first rather increases than diminishes, yet the animal cools as rapidly as if sudden death had occurred. Exactly similar observations have been made on the cutting of the spinal cord and of the pars vagum. The respiratory motions continue for a time, but the oxygen does not meet with those substances with which, in the normal state, it would have combined; because the paralyzed viscera will no longer furnish them. The singular idea that the nerves produce animal heat has obviously arisen from the notion that the inspired oxygen combines with carbon, in the blood itself; in which case the temperature of the body, in the above experiments, certainly, ought not to have sunk. But, as we shall afterwards see, there cannot be a more erroneous conception[1] than this.

'As, by the division of the pneumogastric nerves the motion of the stomach and the secretion of the gastric juice are arrested, and an immediate check is thus given to the process of digestion, so the paralysis of the organs of vital motion in the abdominal viscera

[1] This 'erroneous conception' (that oxygen combined with carbon in the blood) was in fact held by Mayer. (a)

(a) 67, p. 28.

affects the process of respiration. These processes are most intimately connected; and every disturbance of the nervous system or of the nerves of digestion re-acts visibly on the process of respiration.' (a)

Liebig constantly emphasizes, with as much conviction as Bichat showed earlier, that it is in the tissues, bathed by the serum, that one must look for the metamorphoses which are at the base of all growth and reproduction. One of his chapters is in fact headed *The Metamorphosis of Tissues*. (b) He picks up again from Bichat when he emphasizes that the foodstuffs eaten by the animal are dissolved and made soluble, and thus enter the bloodstream. It is the changes that these soluble foodstuffs, and the very tissues themselves, undergo during oxidation and chemical reorganization that produce on the one hand the heat of the animal, and on the other the new material which contributes to growth and reproduction. Moreover, since Liebig was able to give the exact chemical formulae[1] for his organic compounds, and equations showing the changes that occurred when these compounds were oxidized or hydrolyzed, his ideas carried striking conviction.

The influence of Liebig on organic chemistry and physiology was, like that of Wöhler, profound. But there is a paradox here, and their work often is profoundly misunderstood. Read, in his historical survey, *Through Alchemy to Chemistry*, gives the commonly accepted view:

'For a long time it was supposed that none of these plant and animal substances could be made artificially, and that their production was dependent upon life processes and the operation of an imagined "vital force". This belief . . . was undermined in 1828 by an experiment of Woehler, who prepared in the laboratory, without the intervention of any living matter, the substance urea . . . Laboratory synthesis of many other natural organic substances followed, and the idea of a vital force was abandoned.' (c)

Leaving aside for the moment the point that an artificial synthesis of organic compounds can in no way be a crucial factor in determining the presence or absence of a vital force this statement is in any case simply not true.

(a) 59, p. 29. (b) 59, appendix.

[1] *See* footnote on p. 119 above. (c) 79, p. 166.

This particular interpretation of Wöhler's work is of late nineteenth-century origin and arose at a time when chemists and physiologists were looking, one can only suppose, back to a point earlier in the century for heroes who would support their own current and favourite interpretations. This resulted in erroneous assessments of the hero's work, and he and his contemporaries were consequently often credited with beliefs and insights that they did not actually hold. The myth that has grown up around the so-called 'synthesis' of urea is a splendid example of this. McKie (a) has shown that this experiment was actually a transformation of substances rather than a synthesis, and that in this particular case it was Hofman who was Wöhler's prophet. In his obituary of Wöhler (1882) Hofman hails this 'synthesis' as a crucial, significant and epoch-making event for chemistry (and presumably for physiology). In actual fact a study of the documents shows that no particular significance was attached to this experiment by anyone, least of all Wöhler, Berzelius and Liebig. Bertholet (1860) was the first to invoke the aid of organic chemistry to refute vitalism.

But the real paradox is of another kind: it lies in the fact that Liebig and many other chemists, who could already give a perfectly satisfactory explanation of many physiological processes in physical and chemical terms, for all that never abandoned the term 'vital force' as an explanatory factor, when talking about living organisms. The reason for this must be left to the next chapter. For the moment we must just reiterate that, so far as concerns the problem of animal heat, Liebig was the first person to understand the precise *chemical* nature of the process of combustion in the organism. He was able for the first time to give a clear explanation, not only of what it was that was burnt, but also of where it was burnt, how it was burnt and even, to a great extent, why it was burnt. Moreover he managed to relate the processes of respiration and digestion to other physiological processes as well.

So complete was Liebig's success that one may want to

(a) 60, p. 608.

ask what exactly it was that Claude Bernard contributed to this particular issue. The answer is, more precision and more detail in his analysis of all the physiological problems and, above all, the key concept of the 'internal environment'. With this concept he was able at last to explain the basic anomaly of animal heat – the fact that animals seem to disobey physical and chemical laws, and to be free of the external environment in a way no inorganic bodies are. By setting the chemical phenomena of animal heat into their physiological environment in a way that Liebig could never do, he finally dispensed with the concept of a vital force.

We can best study Claude Bernard's work under three headings: first, his view on the site and nature of the combustion which produces animal heat; second, his theories about heat-regulation in the animal; and third, his revolutionary concept of the 'internal environment'.

Bernard often declared that Lavoisier and Laplace's initial experiment was the beginning of scientific physiology, as much by virtue of its methods as of its results. Lavoisier had shown that respiration was a slow combustion: Bernard, in his book on animal heat, takes us through the history of experiments designed to find out where the combustion takes place. He himself was at one stage inclined to Lavoisier's theory, attributing the source of heat to the burning of sugar in the capillary system of the lungs. This, of course, implied a difference in temperature between venous and arterial blood, and Bernard's earliest collaboration with his teacher Magendie (1844) was an attempt to measure the temperature of the blood in different parts of the body. The problem remained in Bernard's mind throughout his life, and a year after succeeding Magendie as Professor of Medicine at the College de France in 1856 he returned to it. (At this time he was unaware that similar work was being done in Germany by Fick and Liebig's son.) Tables of Bernard's results found their way into most contemporary books on physiology. They convinced him of the falsity of Lavoisier's theory: the temperature of the blood in the right side of the heart was always a little higher than that of the blood in the left, whereas according to Lavoisier's theory the position should have been reversed.

> 'Within the depth of the body, venous blood is always warmer than arterial blood. I have found that the hottest venous blood is found in the veins leaving the liver. At the same time there is, in effect, venous blood of two sorts.[1] In the superficial veins of the body the blood is, on the contrary, colder than arterial blood; this can be explained because of the cooling this blood experiences in the vessels where it circulates slowly.' (a)

Bernard had, of course, by this time other reasons for rejecting Lavoisier's hypothesis. To us it may seem fantastic that, as late as 1856, the lungs could still be seriously considered to be the site of the respiratory combustion; but this fact once again reflects the complexity of physiological problems and the near impossibility of devising and performing experiments which were in any way crucial. In order to fill in the background, it is worth going back to 1830 and following out the lines of research that led, through Magendie, to Bernard's interest in animal heat. As we have seen, the important questions at that time still were: Is oxygen absorbed by the blood? Is the carbonic acid gas released from the lungs formed there, or does it already exist as such in the venous blood? Faust and Mitchell believed that the formation of carbonic acid took place in the lung capillaries immediately after the absorption of the oxygen; they could find no trace of carbonic acid gas in the blood. Nor could many other people; but unfortunately the experiments showed widely divergent results. It was up to Liebig to show how little significance these experiments had; attempts to extract gases from the blood by chemical or physical means (he argued) could give no true indication of the actual quantities of gases present in solution.

Meanwhile experiments on these lines continued, and this was one of the threads later to be taken up by Bernard. In

[1] Between 1875 and 1878, when he died, Bernard did many more estimates of the temperature of the blood, now using electric methods. He wished, as he said, to establish the 'topography of the animal temperature' but it was a very difficult study.

(a) 7, *see* esp. ch. 6, p. 108.

1837 Gustav Magnus[1] published a series of estimates of the amount of oxygen and carbonic acid gas in arterial and venous blood obtained both by chemical methods and by direct extraction using an air pump. He found both gases present in both types of blood, though the proportion of carbonic acid gas appeared to be greater in venous than in arterial blood. This result led him to revive the earlier theory, held by Bostock, Lagrange and Hassenfratz:

'It is very probable that oxygen is breathed in and absorbed in the lungs and is carried by the blood around the body, where, in the capillary vessels it contributes to the formation of carbonic acid.'[2] (a)

The lungs, he says, are the site of a purely physical absorption of gases and are not the place where combustion takes place:

'If carbonic acid exists already formed in the venous blood its separation in the lungs takes place by a process like the displacement from a liquid of one dissolved gas by another; and therefore when carbonic acid is evolved a corresponding quantity of oxygen is absorbed in accordance with Mr. Dalton's laws of the absorption of gases into liquids.'[2] (b)

Magnus's results were severely criticized by Gay-Lussac. If oxygen was burnt in the capillary system of the body (he argued) one would expect *arterial* blood to contain this gas, but venous blood should contain only a small amount of it. Yet Magnus's calculations seemed to show the presence of both gases in both types of blood, with only a slight increase in the proportion of carbonic acid gas in venous blood. Gay-Lussac, (c) re-interpreting Magnus's figures, showed that one could even draw the opposite conclusions; that arterial blood contained more carbonic acid[3] gas than the venous did. Magendie was present at the meeting of the Academy of

[1] Liebig did not mention Magnus in his work *Animal Chemistry*. Nevertheless it is very likely that his views on respiration were influenced by Magnus's work, the results of which were published in Poggendorff's *Annalen*, a periodical with which Liebig was closely associated.

(a) 65, p. 93. (b) 65, p. 84. (c) 39, pp. 343–6.

[2] My translation.

[3] Liebig, as we have seen, gives a similar explanation of why more carbonic acid would accumulate in arterial blood after it had been drawn from the animal and left for some time.

Science when these criticisms were presented, and he gave an analysis of the blood, the results of which agreed with those of Gay-Lussac. They decided to do further experiments together on this problem, and Claude Bernard, who by this time was Magendie's assistant, took part in the preparatory work.

For various reasons, the investigation was never completed. One important fact, however, was brought to light, as Bernard tells us:

'If blood is deprived, by the action of a current of hydrogen, of carbonic acid and oxygen . . . and then left for twenty-four hours, an appreciable quantity of carbonic acid is found in the atmosphere of hydrogen, proving that in this case the formation of carbonic acid is not the result of a direct combustion.

'Without doubt today we should be astonished at the objections of Gay-Lussac and Magendie. But it was quite possible to attack Magnus's conclusions because his analyses were by no means conclusive and faultless. He kept the blood too long in the apparatus while under a vacuum. This allowed chemical actions to take place and the two samples of blood to become more and more alike the longer the experiment continued.' (a)

Nevertheless, Bernard continues, since 1844 repeated and accurate analyses of the blood and its gases had confirmed Magnus's conclusions over and over again. One could no longer doubt that oxygen was absorbed in the lungs, and that respiratory combustion took place in the tissues. Bernard himself introduced a new method for analysing the gases in the blood which exploited the tendency of carbon monoxide to displace other gases and itself combine with the haemoglobin. He arrived at this technique indirectly, as a result of his early studies (1846) on carbon monoxide poisoning; and in his book *An Introduction to the Study of Experimental Medicine* (1865) he used the history of this work as an illustration of experimental method in physiology. (b)

Yet, though Bernard accepted that animal heat is produced by a combustion in the tissues, he was careful to draw a distinction between combustion in the chemical sense and 'physiological combustion'.

(a) 7, p. 25. (b) 6, p. 160.

'But with this important modification to the theory of Lavoisier [that combustion occurs in the tissues not in the lungs] ought we to say that there is a direct combustion in the organism and must we conclude that in the general capillaries for example, the oxygen brought there by the arterial blood directly burns the carbon and hydrogen of the blood or the tissues, so that carbonic acid and water are formed, at the same time producing the rise in temperature which is a result of this combustion?

'Today no one denies that the phenomenon of chemical combustion occurs in the organism. Nevertheless even the most eminent chemists do not agree about a precise definition of the nature of this combustion.

'I have said to you many times and I repeated this at the beginning of these lectures that I am one of those who think that the laws of physics and chemistry are not violated in the organism; but on the other hand, I believe that I have demonstrated that though chemical laws are unchanging, chemical processes are variable and are able in certain cases to show such individuality as to become special physiological processes. If I may give my opinion in advance... I would agree in principle that animal heat results from the chemical actions of the organism. I cannot, all the same, accept as proved that it develops from a direct combustion as some chemists have put forward.' (a) [1]

Now Bernard's objections to the idea of 'a direct combustion' in the organism are all physiological, and require us to turn and consider his second contribution to our understanding of the problem of animal heat. Fortunately for us, he again recorded the history of the ideas and the train of researches that led to the discovery of the vasomotor nerves and their part in regulating the temperature of the body.

'About the year 1852, my studies led me to make experiments on the influence of the nervous system on the phenomena of nutrition and temperature regulation. It had been observed in many cases that complex paralyses with their seat in the mixed nerves are followed now by a rise and again by a fall of temperature in the paralysed parts. Now this is how I reasoned, in order to explain this fact, basing myself first on known observations and then on prevailing theories of the phenomena of nutrition and temperature regulation. Paralysis of the nerves, said I, should lead to a cooling of the parts

(a) 7, pp. 26–7. [1] My translation.

by slowing down the phenomena of combustion in the blood since these phenomena are considered as the cause of animal heat. On the other hand, anatomists long ago noticed that the sympathetic nerves especially follow the arteries. So thought I inductively, in a lesion of mixed trunks of nerves, it must be the sympathetic nerves that produce the slowing down of the chemical phenomena in capillary vessels, and their paralysis that then leads to cooling parts. If my hypothesis is true, I went on, it can be verified by severing only the sympathetic, vascular nerves, without loss of either motion or sensation, since the ordinary motor and sensory nerves would still be intact.

'Accordingly I severed the cervical sympathetic nerve in the neck of a rabbit, to control my hypothesis . . . On the basis of a prevailing theory[1] [Brodie's?] and of earlier observation, I had been led, as we have just seen, to make the hypothesis that the temperature should be reduced. Now what happened was exactly the reverse. After severing the cervical sympathetic nerve about the middle of the neck, I immediately saw in the whole of the corresponding side of the rabbit's head a striking hyperactivity in the circulation, accompanied by increase of warmth . . . Thereupon . . . I at once abandoned theories and hypothesis, to observe and study the fact itself, so as to define the experimental conditions as precisely as possible. Today my experiments on the vascular and thermo-regulatory nerves have opened a new path for investigation and are the subject of numerous studies which, I hope, may some day yield important results in physiology and pathology' (a).

His hope, as we know, was more than justified. Further work on the vasomotor reactions led to the view that the difference between warm-blooded and cold-blooded animals was to be found in the possession by the former of controlling mechanisms and not, as had been thought earlier, in the chemical processes of heat production. Yet, during his experiments on the vasomotor nerves, Bernard himself was never fully aware that he was dealing with thermal, as distinct from chemical effects. The hyperactivity he observed in fact involved only the *rate* of blood-flow to the affected part. He felt on the contrary that the nerves not only produced dilations

[1] Experiments on the sympathetic nerve had been done before but Bernard was the first physiologist to notice the rise in temperature. *See* Olmsted's biography of Bernard, p. 208.

(a) 6, p. 168.

of the blood vessels, but also actually speeded up the process of combustion at that point. Conversely,

'The large sympathetic nerve when galvanised produces cold not only by contracting the vessels but also by slowing up and by restraining the chemical processes of nutrition, at the same time.' (a)

He wished to speak of this nerve as producing 'refrigeration', but as always was anxious that this should not be misunderstood:

'When I say that the sympathetic nerve is a chilling nerve I don't want to say that it produces heat and cold in the living organism by nervous action alone, an action mysterious and vital as Brodie and Chaussat had thought. [He is a little unfair to Brodie.] No. The nervous system only produces chilling and warming by its action on the chemical phenomena which accompany the nutrition of the tissues.'[1] (b)

Olmsted has set out very clearly the reasons why Bernard was unable to make a sharp distinction between these thermal and chemical phenomena; that it was difficult for him to believe that mere dilation of the vessels could account for the rise in temperature. He also gives details of Bernard's work on the innervation of the sub-maxillary glands, which showed that the activity of the gland was controlled by two motor nerves; the sympathetic nerve constricting the vessels, the tympanico-lingual dilating them. (c) This work is important here not so much for its details, or for the fields that it opened up, but on account of Bernard's own attitude to his discoveries. He had found yet another regulatory mechanism – one which regulated not only the amount of blood going to a particular organ, but also, as he thought, the chemical processes going on. And it is on this account that he writes:

'To sum up, Lavoisier's theory is nowhere set out as a definitive formulation. There are many spaces, unresolved objections and many unexplained facts in the theory of direct combustion.

'But in its general outlines it is undoubtedly true. The theory cannot be attacked so long as it claims that respiration and the production of

(a) 7, p. 289 *see* esp. chs. 7, 8, 10, 14, 15, for his account of the nervous system in animal heat. (b) 7, p. 290. (c) 70, p. 178. [1] My translation.

heat and physico-chemical facts obeying the ordinary laws of physics and chemistry in general. But the theory is open to serious criticism if it implies that these reactions have nothing special in their processes and take place in the animal just as in a normal situation'. (a)'

This passage illustrates very clearly Bernard's underlying attitude to physiology – and it is this that will be our main concern in the next chapter. He felt that in this connexion one is justified neither in asssuming an exact analogy with ordinary chemical processes, nor in assuming that ordinary laws of physics and chemistry are violated in the organism. Any apparent violation is a result of the complicated compensatory mechanisms whereby living things create for themselves a constant 'internal environment' to buffer the variations in the external world. This, as Olmsted rightly remarks, was Bernard's 'great biological generalization'. (b) And it marks the decisive breakdown of the division, drawn by Bichat and Chaptal, between the destructive physico-chemical world outside the organism, and the constructive physiological world inside it. Using his new concept, Bernard was able to explain away the paradox presented by the living organism – that harmonious and infinitely variable entity, at once seeming to defy the forces of the external world, and at the same time using processes of the same sort to maintain its structure and integrity.

(a) 7, p. 30. (b) 70, p. 252.

[1] My translation.

7

Liebig, Bernard and the Philosophy of Physiology

It is too easy to assume, as biologists often have done, that the sole motive for the use of vitalistic terminology is the concealment of ignorance. One must approach the writings of scientists of the past as one would the writings of those of the present – prepared to believe that they had good reasons for saying what they did. Even if in the end one does not agree with their inferences, one must at least give them consideration before dismissing them. Credit is of course given to Driesch for his work on embryology, and to Bichat for his on histology; but the moment references to 'entelechy' and 'vital properties' appear, these men are dismissed abruptly as vitalists. (Only the fact that Liebig's *Animal Chemistry* seems never to be widely read saves him from the same condemnation.) The reasons for these vitalistic views are rarely given the careful analysis that their authors, as scientists, are entitled to expect – and which is in fact given to the more conventional aspects of their theories. But if one takes the trouble to study these reasons, the issue between mechanists and vitalists no longer seems as clear-cut as it does at first sight. Liebig for one was not a man who would ever have used words or phrases simply to cover up his ignorance. And in this chapter we must analyse the reasons why he felt it necessary to retain the idea of a vital force in his physiological theories, whereas his great contemporary, Claude Bernard, did not. Both these men were brilliant scientists, both were intellectually honest; both accepted physical and chemical methods of experiment as the basis of their studies; and both were determinists. Yet to Liebig vital activities still appeared inexplicable unless one supposed them to be the outcome of a further force beyond the known agencies of physics and chemistry. The chemist was more of a vitalist than the physiologist. Here is a paradox worth examining.

Liebig's book on animal chemistry opens with this categorical definition:

> 'In the animal ovum as well as in the seed of a plant, we recognize a certain remarkable force, the source of growth, or increase in the mass, and of reproduction, or of supply of the matter consumed; a force in a state of rest. By the action of external influences, by impregnation, by the presence of air and moisture, the condition of static equilibrium of this force is disturbed; entering into a state of motion or activity, it exhibits itself in the production of a series of forms, which although occasionally bounded by right lines, are yet widely distinct from geometrical forms, such as we observe in crystallized minerals. This force is called the *vital force*, *vis vitae* or *vitality*.' (a)

Had he said no more, we might have interpreted him as pointing merely to the capacity of ova and seeds to grow in a unique way. His vital force would then have meant no more than the tendency of the ovum – given the right chemical and physical stimulation – to develop into the adult. No explanation is here given of the force: the paragraph as it stands is the description of a potentiality. But if we read further, we find that he does mean more; he believes that there is a unique agency operating in living material, different in nature and manifestation from any other; and moreover, he is prepared to justify this assumption by appeal to the facts he has learnt from his studies in animal chemistry.

> 'If we assume, that all the phenomena exhibited by the organism of plants and animals are to be ascribed to a peculiar cause, different in its manifestations from all other causes which produce motion or change of condition; if, therefore, we regard the vital force as an independent force, then, in the phenomena of organic life, as in all phenomena ascribed to the action of forces, we have the *statics*, that is, the state of equilibrium determined by a resistance and the *dynamics*, of the vital force.' (b)

Like Bichat, Cuvier and others before him, Liebig was struck by the contrast between the rapid decay of organisms at

(a) 59, p. 1. (b) 59, p. 8.

death, and their autonomy while alive. Though in contact with external agents, living creatures were not subject to them, but actually utilized the materials of the external world to maintain their own growth. Such stability of form – like the stability of animal temperature – seemed to Liebig so different from anything in the accepted realms of physics and chemistry, as to be clearly the manifestations of a unique vital force.

'The vital force in living animal tissue appears as a cause of growth in the mass, and of resistance to those external agencies which tend to alter the form, structure, and composition of the substance of the tissue in which the vital energy resides.

'This force further manifests itself as a cause of motion and of change in the form and structure of material substances, by the disturbance and abolition of the state of rest in which those chemical forces exist, by which the elements of the compounds conveyed to the living tissues, in the form of food, are held together . . .

'The vital force causes a decomposition of the constituents of food, and destroys the force of attraction which is continually exerted between their molecules; it alters the direction of the chemical forces in such wise, that the elements of the constituents of food arrange themselves in another form . . . it forces the new compounds to assume forms altogether different from those which are the result of the attraction of cohesion when acting freely, that is, without resistance . . . The vital force is manifested in the form of resistance, inasmuch as by its presence in the living tissues, their elements acquire the power of withstanding the disturbance and change in their form and composition, which external agencies tend to produce; a power which, simply as chemical compounds, they do not possess . . .

'The phenomenon of growth, or increase in the mass, presupposes that the acting vital force is more powerful than the resistance which the chemical force opposes to the decomposition or transformation of the elements of the food . . . (a)

Now this takes us straight back to Bichat and Chaptal. The organism has the capacity to resist the destructive effect of external chemical agencies. Liebig returns to this time and again; death occurs when 'all resistance to the oxidizing power of the atmosphere ceases', when the organs 'have lost the

(a) 59, pp. 186–7.

power of transforming the food into that shape in which it may, by entering into combination with the oxygen of the air, protect the system from its influence . . .' Disease occurs 'when the sum of vital force which tends to neutralize all causes of disturbance [in other words, the resistance offered by the vital force, is weaker than the acting cause of disturbance'.] (a)

If anything, Liebig is even more positive than either Bichat or Chaptal that the cause of this resistance is a vital *force*. There is no question of its reality for him; it is the essential explanatory factor, and the reason why living organisms do not just react chemically with the atmosphere, and decay. Yet Liebig is no copybook vitalist. His vital force, like Bichat's vital properties, has no connexion with the soul or mind, and is directly open to experimental study.

'The higher phenomena of mental existence cannot, in the present state of science, be referred to their proximate, and still less to their ultimate, causes. We only know of them, that they exist; we ascribe them to an immaterial agency, and that, in so far as its manifestations are connected with matter, an agency, entirely distinct from the vital force, *with which it has nothing in common*. We know exactly the mechanism of the eye; but neither anatomy nor chemistry will ever explain how the rays of light act on the consciousness, so as to produce vision.' (b)

Nor is the vital force *immaterial*: it is firmly connected with the matter of the organism. Liebig, of course, is as prepared as Bichat or Claude Bernard to echo Kant's[1] view on the

(a) 59, p. 242. (b) 59, p 7.

[1] Natural science has fixed limits which cannot be passed; and it must always be borne in mind that, with all our discoveries, we shall never know what light, electricity, and magnetism are in their essence, because even of those things which are material, the human intellect has only conceptions. We can ascertain, however, the laws which regulate their motion and rest, because these are manifested in phenomena. In a like manner, the laws of vitality, and of all that disturbs, promotes, or alters it, may certainly be discovered although we shall never learn what life is. Thus the discovery of the laws of gravitation and of the planetary motions led to an entirely new conception of the cause of these phenomena. This conception could not have been formed in all its clearness without a knowledge of the phenomena out of which it was evolved; for, considered by itself, gravity, like light to one born blind, is a mere word, devoid of meaning (Liebig). (c)

For Bichat's views *see* p. 67 above.
For Bernard's views *see* p. 150 below. (c) 59, p. 7.

inevitable limits set to scientific inquiry. But he insists that the science of life or vitality need not be a scrap less reputable than the science of gravity and magnetism.

'The modern science of physiology has left the track of Aristotle. To the eternal advantage of science, and to the benefit of mankind, it no longer invents a *horror vacui*, a *quinta essentia*, in order to furnish credulous hearers with solutions and explanations of phenomena, whose true connexions with others, whose ultimate cause, is still unknown.

'If the vital phenomena be considered as manifestations of a peculiar force, then the effects of this force must be regulated by certain laws, which laws may be investigated; and these laws must be in harmony with the universal laws of resistance and motion, which preserve in their courses the worlds of our own and other systems [physics] and which also, determine changes of form and structure in material bodies [chemistry] altogether independently of the matter in which vital activity appears to reside, or of the form in which vitality is manifested.' (a)

The central problem was to see what analogies he could find between the vital force and the forces of the physical and chemical worlds. Liebig was perfectly clear that this programme meant taking a step which might not prove justified; but from his chemical observations he felt that such analogies were permissible.

'It might appear an unprofitable task to add one more to the innumerable forms under which the human intellect has viewed the nature and essence of that peculiar cause which must be considered as the ultimate source of the phenomena which characterize animal life, were it not that certain conceptions present themselves as necessary deductions from the views of this subject developed in the introduction of the first [chemical] part of this work . . . It must be admitted here, that all these conclusions will lose their force and significance, if it can be proved that the cause of vital activity has nothing in common with other known causes which produce motion or change of form and structure in matter.

'But a comparison of its peculiarities with the *modus operandi* of these other causes, cannot, at all events, fail to be advantageous,

(a) 59, p. 7.

inasmuch as the nature and essence of natural phenomena are recognizable, not by abstraction, but only by comparative observations.' (a)

Now at this time the phenomena of magnetism, electricity, and gravity, as well as those of chemistry – cohesion and affinity – were all expected to find their explanations in terms of 'central forces' of attraction and repulsion. When Liebig introduces the word 'force' into his physiological framework, he treats it as a similar kind of central force, capable of giving rise either to resistance or to motion; so that he is prepared to study it both dynamically and statically.

'The vital force appears as a moving force or cause of motion when it overcomes the chemical forces (cohesion and affinity) which act between the constituents of food, and when it changes the position and place in which their elements occur; it is manifested as a cause of motion in overcoming the chemical attraction of the constituents of food, and is, further, the cause which compels them to combine in a new arrangement, and to assume new forms.

'It is plain that a part of the animal body possessed of vitality, which has therefore the power of overcoming resistance, and of giving motion to the elementary particles of the food, by means of the vital force manifested in itself, must have a momentum of motion, which is nothing else than the measure of the resulting motion or change in form and structure'. (b)

However, in his attempt to show that vital activity has certain properties in common with other natural phenomena, he is obliged to extend his analogy so far that the resulting concept is a curious blend of physical and chemical ideas applied to physiology. It is worth quoting Liebig at length on this point; to show not only how he himself actually conceived the vital force, but also some of the underlying ambiguities which inevitably afflicted him.

'We know that this momentum of motion in the vital force, residing in a living part, may be employed in giving motion to bodies at rest (that is, in causing decomposition, or overcoming resistance), and if the vital force is analogous in its manifestations to other forces, this

(a) 59, p. 185. (b) 59, p. 193.

momentum of motion must be capable of being conveyed or communicated by matters, which in themselves do not destroy its effect by an opposite manifestation of force . . . Motion, by whatever cause produced, cannot in itself be annihilated; it may indeed become inappreciable to the senses, but even when arrested by resistance . . . its effect is not annihilated. The falling stone, by means of the amount of motion acquired in its descent, produces an effect when it reaches the table.

'If we transfer the conception of motion, equilibrium, and resistance, to the chemical forces, which, in their *modus operandi*, approach to the vital force infinitely nearer than gravitation does, we know with the utmost certainty, that they are active only in the case of immediate contact. We know also that the unequal capacity of chemical compounds to offer resistance to external, disturbing influences, to those of heat, or of electricity, which tend to separate their particles, as well as their power of overcoming resistance in other compounds (of causing decomposition), that, in a word, the active force in a compound depends on a certain order or arrangement, in which its elementary particles touch each other.

'. . . If we alter the arrangement of the elements, we are able to separate the constituents of a compound by means of another active body; while the same elements, united in their original order, would have opposed an invincible resistance to the action of the decomposing agent.

'In the same way as two equal inelastic masses, impelled with equal velocity from opposite points, on coming into contact are brought to rest; in the same way, therefore, as two equal and opposite moments of force mutually destroy each other; so may the momentum of force in a chemical compound be destroyed in whole or in part by an equal, or unequal, and opposite momentum of force in a second compound . . .

'As the manifestations of chemical forces (the momentum of force in a chemical compound) seem to depend on a certain order in which the elementary particles are united together, so experience tells us, that the vital phenomena are inseparable from matter; that the manifestations of the vital force in a living part are determined by a certain form of that part, and by a certain arrangement of its elementary particles . . .

'There is nothing to prevent us from considering the vital force as a peculiar property, which is possessed by certain material bodies, and becomes sensible when their elementary particles are combined in a certain arrangement or form.

'This supposition takes from the vital phenomena nothing of their wonderful peculiarity; it may therefore be considered as a resting-point, from which an investigation into these phenomena, and the laws which regulate them, may be commenced; . . .

'A living part acquires, on the above supposition, the capacity of offering and of overcoming resistance, by the combination of its elementary particles in a certain form; and, as long as its form and composition are not destroyed by opposing forces, it must retain its energy uninterrupted and unimpaired.

'When, by the act of manifestation of this energy in a living part, the elements of the food are made to unite in the same form and structure as the living organ possesses, then these elements acquire the same powers . . .

'. . . all matters which serve as food to living organisms are compounds of two or more elements which are kept together by certain chemical forces; if we reflect that in the act of manifestation of force in a living tissue, the elements of the food are made to combine in a new order; it is quite certain that the momentum of force or of motion in the vital force was more powerful than the chemical attraction existing between the elements of the food.

'The chemical force which kept the elements together acted as a resistance, which was overcome by the active vital force.' (a)

Several things must be noted about this passage; first of all, two ambiguities. To begin with, the word 'resistance' appears sometimes to mean actual physical 'pushes and pulls' and sometimes the capacity to slow down or prevent the occurrence of 'physical or chemical processes'. Of course, for anyone thinking in terms of central forces, the 'invincible resistance' of a compound 'to the action of the decomposing agent' would seem explicable only by supposing the decomposing force to be weaker than the binding force. In fact, Liebig himself offers this picture of the state of affairs:

'The hands of a man, who raises, with a rope and simple pulley, 30 lbs. to the height of 100 feet, pass over a space of 100 feet, while his muscular energy furnishes the equilibrium to a pressure of 30 lbs. Were the force which the man could exert not greater than would suffice to keep in equilibrium a pressure of 30 lbs., he would be unable to raise the weight to the height mentioned.' (b)

(a) 59, pp. 194–200. (b) 59, p. 200 footnote.

When he speaks of the chemical force 'active in a compound' as depending on a 'certain order and arrangement of their elementary particles', his precise meaning is again obscure. Sometimes he appears to be referring to the binding forces which hold the constituent atoms together in a molecule, and sometimes to the new forces which the resulting compound molecule becomes capable of exerting, and which give the compound its particular chemical properties. More often than not, he apparently means the latter because his analogy between chemical forces and the vital force implies the following theory. The properties of a chemical compound seem to depend on the configuration of the elementary particles. For instance, the atoms of hydrogen have different properties from those of oxygen, but when an atom of oxygen combines with an atom of hydrogen to form water[1] this configuration of elementary particles produces a new compound. By virtue of its arrangement, the compound has a different set of properties from the original elements; it shows a different type of chemical activity resulting from the *chemical force* which can now be manifested. Yet this same force 'resists' the effects of outside agencies, like heat, which (Liebig supposes) would otherwise tend to break the compound molecule down into its constituent atoms. The same chemical force accounts for the new properties which became apparent only when there is a special configuration of elementary particles, and it also preserves the autonomy of the compounds. Water, for instance, will decompose into its elements only when subjected to a force, for example, an electrical one, which is strong enough to overcome the resistance of chemical affinity. Now, says Liebig, there is nothing to stop us thinking about vital force in the same way; and if we do this, we shall find it a convenient starting point for understanding the laws of vitality. Given a special organic configuration of elementary particles (we suppose) a special and unique type of force will be manifested. Vital activity will then result because of the presence of vital force. This force, though inherent in the material ingredients of the organism, cannot become manifest

[1] *See* note on p. 119.

until they have entered into a certain arrangement or form. But, given the necessary form, the activity will follow; and the vital force will both result from, and maintain it.

Biochemists of the twentieth century may recognize in this statement some of our present ideas about the structure and activity of living material, but with one important difference. We no longer feel it necessary to suppose that a unique force is present in the living material; the other forces of physics and chemistry are enough for us, as they were to be for Claude Bernard. But this fact reflects another, deeper difficulty which faces us when we read Liebig, and which he never overcame. We now have a clear idea of what we mean by the term 'force', and what we require of it in any context. Yet the very word *Kraft* which Liebig uses is profoundly ambiguous. If we look at another passage, we shall see the extent of this difficulty:

'The change of matter, the manifestation of mechanical force, and the absorption of oxygen, are, in the animal body, so closely connected with each other, that we may consider the amount of motion, and the quantity of living tissue transformed, as proportional to the quantity of oxygen inspired and consumed in a given time by the animal. For a certain amount of motion, for a certain proportion of vital force consumed as mechanical force, an equivalent of chemical force is manifested; that is, an equivalent of oxygen enters into combination with the substance of the organ which has lost the vital force; and a corresponding proportion of the substance of the organ is separated from the living tissue in the shape of an oxidized compound.' (a)

Suppose that, for the word 'force', with its precise twentieth-century associations, we here substitute the word 'energy'. The passage at once becomes clear and acceptable. Liebig's word *Kraft* in fact carried the meaning both of our 'force' and of our 'energy'. In 1840, the distinction between the two had not yet been made clear, and the word *Energie* did not appear in scientific papers until some twenty years later. Re-interpreted in this way, this last passage is a clear account of the conservation of energy (*die Erhaltung der Kraft*, as it was first known). In all Liebig's comparisons of gravitational, chemical

(a) 59, p. 211.

and vital 'forces' he kept referring to their conservation. Forces, he says, are never annihilated; their effects alone are different; they are merely transformed from one type to another. This idea of conservation was not entirely new. Carnot had earlier discussed the conservation and reciprocal transformation of heat and mechanical motion. But in this book, Liebig envisages for the first time the reciprocal transformations of forces of all kinds – magnetic, electrical, gravitational, chemical and vital – and tries to give some mathematical expression of their equivalence.

'The act of waste of matter is called the change of matter; it occurs in consequence of the absorption of oxygen into the substance of living parts. This absorption of oxygen occurs only when the resistance which the vital force of living parts opposes to the chemical action of the oxygen is weaker than that chemical action; and this weaker resistance is determined by the abstraction of heat, or by the expenditure in mechanical motions of the available force of living parts . . .

'From the relations between the consumption of oxygen on the one hand, and the change of matter and development on the other, the following general rules may be deduced.

'. . . The sum of force available for mechanical purposes must be equal to the sum of the vital forces of all tissues adapted to the change of matter.

'If, in equal times, unequal quantities of oxygen are consumed, the result is obvious, in an unequal amount of heat liberated, and of mechanical force.' (a)

Liebig lists twenty-one general rules relating the vital force to other unique forces. Each rule expresses a direct quantitative relationship. Despite his references to the uniqueness of the vital force, the whole scheme is as exact and determined as the electromagnetic and gravitational theories with which he liked to compare it. Indeed, though he has no units for his vital force, he is nevertheless able to give its mathematical equivalent in terms of mechanical force.

'That condition of the body which is called *health* includes the conception of an equilibrium among all the causes of waste and of

(a) 59, p. 231.

supply; and thus animal life is recognized as the mutual action of both; and appears as an alternating destruction and restoration of the state of equilibrium.

'In regard to its absolute amount, the waste and supply of matter is, in the different periods of life, unequal, but, in the state of health, the available vital force must always be considered as a constant quantity, corresponding to the sum of living particles.' (a)

He even gives a table, showing how the vital force is expended at different ages:

	'Force expended in mechanical effects	Force expended in formation of new parts
In the adult	17	7
In the infant	4	20
In the old man	20	4

'In the adult, a perfect equilibrium takes place between waste and supply; in the old man and in the infant, waste and supply are not in equilibrium. If we make the consumption of force in the 17 waking hours equal to that required for the restoration of the equilibrium during sleep = 100 = 17 waking hours = 7 hours of sleep, we obtain the following proportions. The mechanical effects are to those in the shape of formation of new parts

'In the adult man = 100 : 100
In the infant = 25 : 250 (24 : 286)
In the old man = 125 : 50 (118 : 43)'

Or the increase of mass to the diminution of waste:

'In the adult man = 100 : 100
In the infant = 100 : 10 (9)
In the old man = 100 : 250 (274) (b)'

This was both new and visionary. The principle of the conservation of energy is usually credited to Mayer,[1] who had close associations with Liebig and Poggendorf. Reading Liebig, one wonders whether an injustice has not been done.

(a) 59, p. 233. (b) 59, p. 237–8.

[1] But, *see* Kuhn's excellent paper (c) on the simultaneous discovery of this concept. Kuhn does not however make the distinction between '*Kraft*' and '*Energie*'. (c) 48, p. 321.

If one compares the relative clarity with which Liebig sets out his ideas on the subject, and his attempts to give them mathematical expression, with the diffuse metaphysics of Mayer's first paper (a) published in the same year, one cannot help feeling that Mayer's later precision owed a tremendous amount to Liebig's influence. Anyone who goes back and reads Mayer's 1842 paper in the original will understand why it was at first refused publication. Perhaps it is only Liebig's preoccupation with vital force that has deprived him of some of the credit given to his protégé Mayer.

It is only in a very modified sense, then, that Liebig can be called a vitalist. Certainly his experimental technique is as mechanistic as anyone could ask. Only by comparing the organism's behaviour with the behaviour of inanimate things, he feels, can one get a clearer understanding of what is going on within it – and this means studying it by physico-chemical methods. Moreover, many of the processes taking place in an animal's body are obviously chemical and have to be studied as such; they are as strictly determined as chemical processes going on outside the organism. Here Liebig differs radically from Bichat. He explains the variability in the observed behaviour of organisms not as a sign that they are exempt from the determinism of nature, but as a consequence of the varying conditions of environment and the complex interconnexions of the organs inside them. He utterly rejects Bichat's insistence on the essential invariability of vital processes, and in this he is far closer in spirit to Claude Bernard.

Yet Liebig must be called a vitalist despite his impeccably mechanist methodology. He claimed to observe an 'agency' in living things which had no counterpart in non-living things; and he did not see how without this concept vital activities could possibly be explained. It seems to have been his very preoccupation with physical and chemical ideas that at this point led him off down a false trail. The uniqueness of the activities to be accounted for was and is beyond question. Some new concept was needed, if the fundamental principles

(a) 66.

of physics and chemistry were to be applied in this new field. What form should this new concept take? For a well-brought-up nineteenth-century chemist, accustomed to regarding *force* as the ultimate explanation of all things, it must have seemed the most natural thing in the world to suppose that one more force was here at work, and that by studying this, the general theory enunciated by Newton could be applied to vitality – as it had been earlier, for example, to magnetism.

In making this supposition, Liebig was in fact following the best modern maxims. If the facts of physiology call for new concepts (said Charles Vernon in 1958), these must be framed as extensions of existing *chemical* concepts. (a) For example, he admits that certain important ideas of twentieth-century biochemistry, such as that of the 'high-energy phosphate bond', cannot be reduced to simple chemical terms; they refer essentially to *organized* chemical reactions taking place within the ordered structure of the living body. The problem is to introduce into our theories concepts effective for physiology, which will at the same time 'escape the charge of vitalism'. (b) We can escape such a charge, he concludes, only if these new concepts are 'clearly reducible to chemical concepts'. Yet we started, surely, from the recognition that existing chemical ideas alone were not capable of covering all the facts in question; and only anti-vitalist prejudice, it seems, prevents Vernon from allowing physiology to have autonomous concepts of its own.

In the story which concerns us here, it was the opposite attitude which proved effective. Liebig, striving to extend the ideas of physics as they stood into the realm of physiology, was diverted into an unprofitable form of vitalism; Bernard avoided this trap by doing the very thing Vernon deplores, and in the process framed for the first time the central concepts of an autonomous scientific physiology. Throughout his life Bernard made it quite clear that physiology was a science allied to physics and chemistry, and not accessory to them. Physiology needed to develop theories and concepts of its own. In his classic book, *An Introduction to the Study of Experi-*

(a) 84, p. 579. (b) 84, p. 579.

mental Medicine, published in 1865, (a) he insists that the science of physiology must find its explanation through 'physics and chemistry worked out in the special field of life'. If we confuse the problems and points of view of the various sciences we risk being led astray in our investigations:

'We have seen, and we still often see, chemists and physicists who, instead of confining themselves to the demand that living bodies furnish them suitable means and arguments to establish certain principles of their own sciences, try to absorb physiology and reduce it to simple physico-chemical phenomena. They offer explanations or systems of life which tempt us at times by their false simplicity, but which harm biological science in every case, by bringing in false guidance and inaccuracy which it then takes long to dispel. In a word, biology has its own problems and its definite point of view; it borrows from other sciences only their help and methods, not their theories'. (b)

Perhaps Bernard might have levelled this charge against Liebig, who came to physiology as a chemist – and from the outside. Bernard, though always insisting on the same mechanistic methodology as Liebig, approached his problems always from the point of view of a physiologist – from within the subject. He never once lost sight of the organism as an integrated whole, interacting with its environment; whereas Liebig, with his eyes focused closely on the bio-chemical processes within it, seems farther away from his object of study. Thus it was that Bernard, from within physiology, was able to dispense with the idea of a vital force and to recognize what was needed instead – namely, a clear understanding of that complex system which he now christened the 'internal environment'.

The book which marks the climax to Bernard's career was written when he was fifty-two. For twenty-six years he had worked in a school where the emphasis was on experiment as the main tool of physiological investigation; where it was taken for granted that hypotheses (which Magendie so disliked and never employed if he could help) were to be subordinate to the facts of observation. Magendie's antivitalist prejudices

(a) 6. (b) 6, p. 95.

had led him to reject the phrase 'vital force' whenever he could – solely because it was hypothetical. (As we have seen he was not able for all that to dispense entirely with the *idea.*) It was left to Claude Bernard to redress, in some measure, the swing of the pendulum; to show the relation between hypothesis and experiment in physiology. It was not yet universally clear what physiology was really concerned with, and what was its relation to the other sciences; and while the appeal to vital forces as *explanatory* factors was still so widespread, Bernard never lost any opportunity to emphasize the proper limits of physiological speculation.

Two things, he felt, were fundamental. First, anything metaphysical must be left out of physiology; and second, physiologists must recognize the same methodological limits to their studies as physicists admit to theirs. Here are some of his final words on the subject:

'In short, the idea of a metaphysical force of development by which we might try to characterize life, is useless to science because being outside the forces of physics, it cannot possibly exercise any influence on them. It is here necessary to separate the metaphysical world from the world of physical phenomena . . . Leibniz has expressed this demarcation in the words which we recalled at the beginning of this study . . . Science today hallows the distinction.

'Even if we could define life by a special metaphysical concept, it is no less true that mechanical, physical and chemical forces are the only effective agents in the living organism and that the physiologist has to take into account only their action.

'We will say with Descartes: one thinks metaphysically but one lives and acts physically.' (a)

The limits which Bernard set to physiology were a consequence of his definition of it, as 'the science whose object it is to study the phenomena of living bodies and to determine the material conditions in which they appear.'

'Neither physiologists nor physicians need imagine it their task to seek the cause of life or the essence of disease. That would be entirely wasting one's time in pursuing a phantom. The words, life, death,

(a) 8, p. 211.

health, disease have no objective reality. We must imitate the physicists in this matter and say, as Newton said of gravitation . . . "But the first cause which makes these bodies fall is utterly unknown. To picture the phenomena to our minds, we may say that the bodies fall as if there were a force of attraction towards the centre of the earth . . . But the force of attraction does not exist, we do not see it; it is merely a word used to abbreviate a speech".' (a)

Similarly, to use the words 'vital force' in any other way than as a physicist would use the word 'gravity' – that is descriptively – was, Bernard felt, to overstep the limits of one's method. But if vitalists would recognize and restrict the use of these words to the facts of observation then he, too, would agree with them.

'When a physiologist calls in vital force or life, he does not see it; he merely pronounces a word; only the vital phenomenon exists with its material conditions; that is the one thing he can study and know . . .

'. . . I should agree with the vitalists if they would simply recognize that living beings exhibit phenomena peculiar to themselves and unknown in inorganic nature. I admit, indeed, that manifestations of life cannot be wholly elucidated by the physico-chemical phenomena known in inorganic nature . . . I will simply say that if vital phenomena differ from those of inorganic bodies in complexity and appearance, this difference obtains only by virtue of determined or determinable conditions proper to themselves. So if the science of life must differ from all others in explanation and in special laws, they are not set apart by scientific method.' (b)

Determinism is the basis of all science; at once its axiom and its postulate. To create a science of physiology, it is necessary – and sufficient – to show that physiological phenomena are causally determined. If, instead, one supposes that there is a force in living things in opposition to physico-chemical forces, dominating all phenomena of life, and subjecting them to entirely separate laws, then there can be no science of biology. His whole aim was to show that determinism applied to vital phenomena as much as to the realms of physics and chemistry, and that the experimental methods of all these sciences must be the same.

(a) 6, p. 67. (b) 6, p. 69.

'We should either have to recognize that determinism is impossible in the phenomena of life, and this would be simply denying biological science; or else we should have to acknowledge that vital force must be studied by special methods and that the science of life must rest on different principles from the science of inorganic bodies . . . I propose, therefore, to prove that the science of vital phenomena must have the same foundations as the science of the phenomena of inorganic bodies, and that there is no difference in this respect between the principles of biological science and those of physico-chemical science. Indeed, as we have already said, the goal which the experimental method sets itself is everywhere the same; it consists in connecting natural phenomena with their necessary conditions or their immediate causes.'[1] (a)

As Bernard points out, the case for independence in physiology is usually made plausible by the restriction of discussion to the higher and more complex animals. This is artificial. In the simpler forms of life even the appearance of independence is not found; their vital properties are never apparent without the right environmental conditions, such as heat, light and moisture. So, he concludes, these influences

'. . . are exactly the same as those which produce, accelerate or retard manifestations of physico-chemical phenomena in inorganic bodies, so that instead of following the example of the vitalists in seeing a kind of opposition or incompatibility between the conditions of vital manifestations, we must note, on the contrary, in these two orders of phenomena a complete parallism and a direct and necessary relation. Only in warm-blooded animals do the conditions of the organism and those of the surrounding environment seem to be independent; . . . an inner force seems to join to combat with these influences and in spite of them to maintain the vital forces in equilibrium.

'But fundamentally it is nothing of the sort; and the semblance depends simply on the fact that, by the more completely protective

[1] Bernard gave credit to Bichat for this insight. 'Bichat by his genius realized that the cause of vital phenomena must be sought . . . in the properties of master within which the phenomena are manifested. Certainly Bichat did not define vital properties . . . his genius . . . lay . . . in being the first to express this general idea, at once illuminating and fruitful, that in physiology as in physics, the phenomena are to be related back to the properties as their cause.' *My translation.* (b)

(a) 6, p. 60. (b) 8, p. 158.

mechanism which we shall have occasion to study, the warm-blooded animals' internal environment comes less easily into equilibrium with the external cosmic environment. External influences, therefore, bring about changes and disturbances in the intensity of organic functions only in so far as the protective system of the organism's internal environment becomes insufficient in given conditions.' (a)

Instead of treating the organism as a collection of separate organs maintained and harmonized by the action of a special internal force as Liebig had done, we are to see these organs as developing within, and linked together by a special 'internal environment'. This was the new conception with the help of which, Bernard argued, the realm of physics and chemistry could be extended to include the whole organism; and which would at the same time account for the remarkable integration and stability of higher animals.

'The general cosmic environment is common to living and to inorganic bodies; but the inner environment created by an organism is special to each living being. Now, here is the true physiological environment; this it is which the physiologists and physicians should study and know, for by its means they can act on the histological units which are the only effective agents in vital phenomena. Nevertheless, though so deeply seated, these units are in communication with the outer world; they still live in the conditions of the outer environment perfected and regulated by the play of the organism.' (b)

The internal environment is formed by the intracellular fluids, the serum of the blood, and the circulating liquids of the body. As the organism becomes more and more complex, so this environment becomes more and more isolated from the outside world, and the differences between the simpler and more complex creatures are simply differences in this degree of 'isolation and protection'. Even the normal physiological behaviour of the higher animals is conditional to the constancy of their internal environments:

'Vital manifestations in animals vary only because the physico-chemical conditions of their internal environments vary; thus a mammal, whose blood has been chilled either by natural hibernation

(a) 6, p. 62. (b) 6, p. 76.

or by certain lesions of the nervous system, closely resembles a really cold-blooded animal in the properties of its tissues.' (a)

Up to a point, we can now see, Chaptal had been right.[1] The internal parts of the body *are* withdrawn from and unaffected by external influences – not because of a fundamental opposition of kind, which exempts them from the laws of chemistry, but because the mechanisms regulating the internal environment act as a controlling buffer. Thus the physical characters of the body-fluids are kept constant.

The phenomenon of animal heat is a perfect illustration of the general point. To explain how an animal keeps its temperature constant in the face of external variations, one need not assume that there are variable vital laws – blowing now hot, now cold – but one must recognize a system of checks and balances. The vasomotor reactions which Bernard himself had discovered, the mechanism by which the sugar content of the blood is maintained, and so on – these all counteract any influences tending to vary the body temperature. Yet this system of checks and balances is itself strictly determined, and follows the laws of physics and chemistry. A superficial inspection of a thermostatically controlled electric iron might lead the uninformed to suppose that the iron contained an agency capable of acting directly against external temperature changes – an agency which was quite different from other agents. Yet all it depends on is a regulatory device, by which more or less heat is allowed to be produced according to the external conditions, and this operates by a perfectly definite physical process.

The immediate task of physiology, then, was to study the physico-chemical interaction between the fundamental organic units – the cells – and the internal environment surrounding them. If we adopt this approach we shall find that we have no need to bring into our account forces opposed to those of physics and chemistry.

Each of the body's cells is in certain respects autonomous, but the activities of all the cells are kept in step, partly by the

(a) 6, p. 64. 1 *See* passages quoted above, p. 64.

integrative action of the nervous system – which Brodie had suspected and Sherrington was to study later – and partly by the vasomotor and other mechanisms which were Bernard's special subject-matter. Here, in the internal environment, were the first of the integrating factors which had mystified previous physiologists. And they proved on examination to be as subject to the laws of physics and chemistry as the external, inorganic environment, whose properties had been understood for so much longer.

8

The Resolution of the Mechanist-Vitalist Dispute

THE story we have been telling has three strands. First, there is the gradual emergence of a scientific physiology, with its own concepts, methods of inquiry, and types of explanation. Second, there are the philosophical debates which developed in the course of this emergence, concerning the legitimacy of using (or rejecting) experimental and quantitative procedures in physiology, or admitting physico-chemical (or non-physico-chemical) concepts into physiological explanations. Finally, as an illustration of the practical applications of these apparently somewhat abstract issues, there is the progressive clarification of men's ideas about the origin and control of animal heat.

It was no accident (I suggest) that the manner in which living things maintain and control their temperature remained in part mysterious until the moment when physiology acquired a definitive method. For so long as men were able only to analyse the processes going on in individual *organs* in physical and chemical terms and had not found a way of extending this method of analysis to functions of the whole organism, these overall functions – regulatory or otherwise – were bound to remain imperfectly understood.

It was here that Claude Bernard's genius showed itself. He saw every physiological problem as involving physical and chemical relations between three things – the organs directly concerned, the organism as a whole, and the environments of both. He was at least as skilful an experimenter as his master Magendie, and used to say – what Magendie would have applauded – that when going into the laboratory to do an experiment one should leave one's imagination outside with one's overcoat. Unlike Magendie, however, he saw how necessary it was, if one were fully to appreciate the implications of one's experimental results, to put this imagination on again when one left.

Certainly it took imaginative insight to see the potentialities of the concept of the internal environment, which (as we saw) was Bernard's most original contribution to physiological theory. For here at last was an idea providing terms in which one could discuss the physico-chemical properties and activities of the complete, integrated organism. There were plenty of physiological problems which remained unexplained even then – and still do, for that matter – particularly in the fields of embryology and regeneration. But *control* had for so long been regarded as the specific function of the vital principle that a break-through on this front was of great importance.

Bernard's picture of the organism also made it possible to recognize at last certain crucial ambiguities in the traditional debate between vitalists and mechanists. For (he rightly insisted) there are in organisms *processes*[1] of a unique kind, unlike any observed by chemists in the inorganic world; but these processes are not necessarily inexplicable in terms of the same theoretical principles as apply in the inorganic world. It is accordingly unhelpful to state the central question dividing mechanists from vitalists by asking whether organisms obey the same *laws* as non-living things. The term 'laws' covers so many different ideas in science that the only correct answer to this question is 'yes and no'.

Physicists speak at the experimental level, for instance, of 'Newton's Law of Cooling'; and the very problem of animal heat arises only because – as a matter of direct observation – warm-blooded animals fail to conform to this law. If this is a typical example of a physical law, then obviously organisms can evade them; the vitalists win without any argument. But alongside this are Newton's Laws of Motion and Law of Gravitation, and these have a very different status. Direct observation of rolling balls or climbing aeroplanes, without

[1] See his important distinction between 'laws' and 'processes', quoted on p. 131.

theoretical interpretation, is not relevant to their truth or applicability; and the existence of vital phenomena which at first glance seem to be at variance with theoretical ideas of physics and chemistry need be no more conclusive. Claude Bernard himself makes use of this very comparison:

> 'If we limit ourselves to the survey of total phenomena visible from without, we may falsely believe that a force in living beings violates the physico-chemical laws of the general cosmic environment, just as an untaught man might believe that some special force in a machine, rising in the air or running along the ground, violated the laws of gravitation.' (a)

There may, at the level of observation, be unique vital activities and processes – as descriptive vitalists never tire of pointing out; but the question of whether the mechanisms behind these activities involve unique substances, or forces, or agencies – as explanatory vitalists have always maintained – is a quite different one.

A failure to recognize this distinction has led to an over-simplification in attitude towards the mechanist-vitalist dispute, both among those who have directly participated in it, and among the historians who have recorded it; and it continues to give rise to excessive dogmatism on both sides. For there will always be something unique to be said about biological systems and happenings. The laws of physics and chemistry by themselves say nothing about the configurations in which matter *must* occur, or the sorts of complex process which will *in fact* be found: these laws are all conditional, enabling us to work out what consequences we may expect *if* particular bodies have such-and-such a physical arrangement or chemical structure. It is for physiology to establish how organisms are in fact constituted, and what processes go on in them. And at its own level, biology may very well be found to have laws of its own, which are irrelevant to anything in the inorganic world – Mendel's Laws of Inheritance provide an obvious example. Yet nobody supposes that the validity of Mendel's Laws (for instance) implies any resistance

(a) 6, p. 63.

by the germ-cells to the operation of physical and chemical influences. The existence of biological laws at one level is quite compatible with biological happenings being entirely covered – at another level – by our physical and chemical theories.

In the history of the dispute as we have followed it in this book, it was never possible to say for certain that *no* special force, substance, or type of energy would be found necessary in physiology. It had in each case to be shown quite clearly that the substances, forces and energies already known from the study of inorganic nature were sufficient for the purpose; and this often took a lot of doing. The labour involved may seem in retrospect to have been wasted. But if Karl Popper (a) is right in regarding the falsification of rival hypotheses as the strongest way in which the adequacy of a theory can be established, it was in fact indispensable. And until it had been done, some people quite reasonably hesitated to apply physical and chemical laws and ideas indiscriminately to their study of living creatures.

Claude Bernard himself always recognized the danger of oversimplifying the problems of physiology. Though he frequently applauded Lavoisier and Laplace for their pioneer excursion into experimental biochemistry, he could (on his own principles) never consistently have agreed with his biographer, Mrs. E. H. Olmsted, that 'their work on respiration and animal heat had established that the *same physical and chemical laws hold for animate as for inanimate bodies*.'[1] (b) The whole issue, as he repeatedly insisted, was far more complicated than that. And on this very subject he quoted and endorsed a passage by Regnault and Reiser about 'the calculations used to establish the theory of animal heat', in which they describe the phenomenon as 'much too complex for possible calculation of the heat from the quantity of oxygen consumed,'[2] ending with the statement that

'It was therefore by a fortuitous circumstance in the experiments of Lavoisier, Dulong and Despretz, that the quantity of heat liberated

(a) 76a. [1] My italics. (b) 73, p. 155.

[2] Cf. Liebig's remarks, p. 121 above.

by an animal was found to be about equal to what the carbon (contained in the carbonic acid produced) and the hydrogen would give off in burning . . .' (a)

Before the relations between physics, chemistry and physiology could be clearly demonstrated and understood, a great deal more than Lavoisier and Laplace were ever able to do had to be accomplished.

It would be misleading to suggest that Claude Bernard himself was completely out of the wood, either physiologically or philosophically. By careful attention to the distinction between processes, laws, and theories, he was able in practice to reconcile his sense of the uniqueness of living things with his belief in physico-chemical determinism. But he could not entirely pull himself above the dispute with which he had grown up; and the errors into which uncritical vitalists were liable to fall always shocked him profoundly.

'The habit of vitalistic explanation makes us credulous and promotes the introduction of erroneous or absurd data into science. Thus, quite recently I was consulted by an honorable and much respected practising physician who asked my opinion of a most unusual case, of which, he said, he was very sure, because he had taken all precautions necessary to observing it well: here was a woman in good health except for a few nervous anomalies, who had neither eaten or drunk anything for several years. Evidently the physician was persuaded that vital force is capable of anything, so that he sought no other explanation. The slightest idea of science, however, and the simplest notions of physiology, would have been enough to undeceive him, by showing that his statement very nearly amounted to saying that a candle can go on shining and burning for several years without growing any shorter.' (b)

In biology itself, the problem of development continued to baffle him. He never clearly declared his views on the Darwinian doctrine, but his recorded references to the subject all sound more like Cuvier than Darwin. Embryological development, however, was a subject to which he returned again and again; and here we find the old mechanist-vitalist

(a) 6, p. 133. (b) 6, p. 202.

dispute continuing in his own person. For on the one hand, all the material changes taking place in the embryo must (he is sure) be 'purely chemical in nature'. Yet on the other hand, he finds himself unable to characterize the facts of development and regeneration without bringing into his account references to a 'developmental force' in the egg:

'The primary essence of life is a developing organic force, the force which constituted the mediating nature of Hippocrates and the *archeus faber* of Van Helmont. But whatever our idea of the nature of this force, it is always exhibited concurrently and parallel with the physico-chemical conditions proper to vital phenomena . . . (a)

'When a chicken develops in an egg, the formation of the animal body as a grouping of chemical elements is not what essentially distinguishes the vital force. This grouping takes place only according to laws which govern the chemico-physical properties of matter; but the guiding idea of the vital evolution is essentially of the domain of life and belongs neither to chemistry nor to physics nor to anything else. In every living germ is a creative idea which develops and exhibits itself through organization. As long as a living being persists, it remains under the influence of this same creative vital force, and death comes when it can no longer express itself; here as everywhere, everything is derived from the idea which alone creates and guides; physico-chemical means of expression are common to all natural phenomena and remain mingled, pell-mell, like the letters of the alphabet in a box, till a force goes to fetch them, to express the most varied thoughts and mechanisms. This same vital idea preserves beings, by reconstructing the living parts disorganized by exercise or destroyed by accidents or diseases.' (b)

He is well aware, of course, that this conception of a creative vital force may appear to be the 'last refuge of vitalism'. But since he intends the notion to be understood, not in a scientific sense, but in terms borrowed from Kantian metaphysics, he thinks he can defend it against the objections which a rigid mechanist might be expected to raise.[1]

In the period since Claude Bernard, the focus of interest in

(a) 6, p. 92. (b) 6, p. 93.

[1] See quotation on page 152.

physiology and biochemistry has shifted to other fields, notably those of genetics, the mechanisms of the nerves and the brain (which Brodie and Berzelius had thought would forever escape mechanistic explanation), as well as embryological development. In each case, we find some of the same divisions of opinion that arose earlier over temperature-control in warm-blooded animals. And it has become progressively clearer that dogmatic assertions from either side of the fence do nothing to solve the essential difficulties.

As a matter of simple description and observation, the intellectual capacities of the higher animals, the development of the embryo, and the self-duplication of the gene and the virus are undeniable and have no counterpart in simpler systems. Those who wish to insist on these facts can do so, without having to abandon the attempt to understand the mechanisms through which the functions operate. Conversely, though we may confidently suppose that physiologists and biochemists will one day uncover these *mechanisms* in all their complexity, they will have done nothing thereby to *explain away* the special character of the associated functions. For it is these functions and activities that are our starting-point, and pose our problem.

It was for this reason that Bernard in his later years called himself a physical vitalist, and renounced both of 'the two schools which make vital phenomena something absolutely distinct from physico-chemical phenomena or something completely identical with them'. (a) Any explanation of natural phenomena which does not include an account of the physico-chemical mechanisms involved will never be accepted as complete; but the mechanisms have to be adapted to the phenomena. Biophysics and biochemistry need introduce no new forces, energies or elements – though they may indeed discover new types of molecular configuration. What makes them the subjects they are is the fact that they are physics and chemistry 'carried out in the *special field of life*'. For a complete picture of living creatures and their activities, we shall always

(a) 73, p. 124; *see also* reference 8.

require an alliance of mechanisms derived from physics and chemistry and concepts proper to biology. In this sense there would seem to be no need for a *direct* conflict between mechanists and vitalists.

Yet the dispute continues to arise – even now.[1] That it does so is, in my opinion, because the dispute is fundamentally one about intellectual *policy*. As such, it displays most of the characteristics of a political dispute and similar sorts of temperamental difficulties separate the major parties. At any one time this question arises: can our existing ideas, with a little modification, meet the demands of contemporary problems, or is some new reorganization or concept called for? As each new generation faces a new situation so the precise line of battle shifts. In consequence two things tend to happen.

First, each new change or concept appears to solve the crucial problems, only for a new situation to arise a generation later, creating new problems and calling for fresh intellectual decisions. For instance, the mechanist-vitalist dispute has arisen in biology in the last 160 years successively in the fields of biochemistry, neurophysiology and embryology.

The second thing that results from the form of this dispute is that ideas which appear to be radical in one generation come to be accepted as conservative in the next. Bernard's essentially *biological* concept of the 'internal environment' is by now quietly appropriated into the most respectable mechanistic theory. (It is not only political parties that steal each others clothes!)

If the story of this dispute has any moral it is this: Science has its own most effective ways of dealing with frankly impossible ideas. Leaving these aside, one cannot legislate in advance as to the types and forms of concepts that are going to prove necessary and most useful in the years to come. It is instructive to recognize that chemists may be forced to borrow ideas from biology instead of *vice versa*. Consider for instance Calvin's suggestion that 'natural selection' may be relevant to an understanding of the chemical changes

[1] See a symposium: *The Historical Development of Physiological Thought* (ed. Brooks and Cranefield), New York, 1959.

which preceded the origin of life on this earth.) Watching the two-way flow of ideas by which different sciences have always cross-fertilized each other – and continue to do so – the philosophically-minded historian of science can hardly take sides in a dispute as we have been considering. He will probably decide, instead, to keep one foot firmly planted on each side of the fence.

Biographical Bibliography[1]

ABERNETHY, J., 1764–1831. Surgeon at St. Bartholomew's Hospital, London. According to his pupil Brodie, and others, he was a brilliant lecturer and teacher as well as an eminent surgeon.

1. *An Inquiry into Mr. Hunter's Theory of Life*, London, 1814.

ALLEN, W., 1770–1843, and Pepys, W. H., 1775–1856. William Allen was a notable chemist, founder of the firm of Allen and Hanbury, Fellow of the Royal Society (1809), and lecturer at Guy's Hospital and the Royal Institution. A member of the Society of Friends, he became one of the prominent humanitarians of his time (*D.N.B.*, and *Through a city archway: the story of Allen and Hanburys, 1715–1954*, by D. Chapman Huston and E. C. Cripps, London, 1954).
William Pepys, Fellow of the Royal Society (1808) and a President of the Royal Institution (1816), was a philosophic instrument maker with an extraordinary flair for the invention of scientific apparatus.

2. 'On Respiration.' *Philosophical Transactions of the Royal Society*, London, 1809, page 404.

3. 'On the Changes produced in the atmospheric Air, and Oxygen Gas by Respiration.' *Nicholson's Journal*, 22, 1809.

ARISTOTLE, 384 B.C.–322 B.C.

4. 'On Respiration.' *On the Soul, Parva Naturalia, On Breath.* Translated by W. S. Hett, Loeb Classical Library, London, 1957, p. 431.

BERNARD, C., 1813–78.

5. 'Notice sur M. Magendie', *Leçons d'Ouverture du Cours de Médecine au College de France*, 1836, p. 7.

6. *Introduction à l'étude de la Médecine expérimentale*, with a memoir of Bernard by Paul Bert. Paris 1865. Translated into English by Copley Greene, 1927. Reprinted by Dover, 1957.

[1] Biographical detail is given only of those doctors, important for this book, who are not so well known and who are not usually referred to in textbooks on the history of science.

7. *Leçons sur la Chaleur Animale*, Paris, 1876.

8. 'Définition de la Vie' (1875), reprinted in *La Science Expérimentale*, *Paris*, 1878.

BERZELIUS, J. J., 1779–1848.

9. *A View of the Progress and Present State of Animal Chemistry*, translated by Gustav Brunnmark, London, 1813.

BICHAT, M. F. X., 1771–1802. French Physiologist and Doctor.

10. *Anatomie Générale appliquée à la Physiologie et Médecine*. Paris, 1801. Translated by C. Coffyn.

11. *Recherches Physiologiques sur la Vie et la Mort*, 3rd Paris, 1805. Reprinted in the collection: *Les Maîtres de la Pensée Scientifique*, Paris, 1955.

BLUMENBACH, J. F., 1752–1840. Comparative anatomist and embryologist at Goettingen.

12. *The Institutions of Physiology*, 3rd edition, translated by John Elliotson, London, 1817.

BLACK, J., 1728–99.

13. *Lectures on the Elements of Chemistry*, published from his manuscripts by John Robison, Edinburgh, 1803.

BOSTOCK, John, *the younger*, 1773–1846. Forsook the practice of medicine for chemistry, physiology, and general science; was president of the Geological Society, and first a Fellow of the Royal Society and then Vice-President (1832). Famous as the author of the first detailed description of hay fever (*D.N.B.*).

14. *An Essay on Respiration*, Liverpool, 1804.

15. 'Remarks on Mr. Ellis's Theory of Respiration', *Edinburgh Medical Journal*, *4*, 1808, p. 159.

16. *An Elementary System of physiology*, 3rd edition, London, 1836.

BRODIE, Sir Benjamin C., *Bart*, *the elder*, 1783–1862. Fellow of the Royal Society (1810), and president (1858–61); President of the Royal College of Surgeons (1844). A tireless worker, he did valuable work in physiology before he turned to surgery. Brodie's abscess, Brodie's disease, Brodie's ligament, Brodie's tumour, are all named after him (*D.N.B.*).

17. 'Some Physiological Researches respecting the Influence of the Brain on the Action of the Heart, and on the Generation of Animal Heat.' *Philosophical Transactions of the Royal Society*, 1811. p. 36. (The paper was actually read on 20 December, 1810.)

18. 'Further Experiments and Observations on the Influence of the Brain on the Generation of Animal Heat.' *Philosophical Transactions of the Royal Society*, London, 1812, p. 378.

CARLISLE, Sir Anthony, 1768–1840. F.R.S. (1800), President of the Royal College of Surgeons (1829). Described as 'neither a brilliant anatomist nor physiologist but . . . a fairly good surgeon' (*D.N.B.*). He was a voluminous writer, and in his youth took part in important researches in voltaic electricity.

19. 'The Croonian Lecture of Muscular Motion', *Philosophical Transactions of the Royal Society*, London, 1805, p. 1.

CHAPTAL, J. A., 1756–1832. French chemist at Montpellier, responsible for naming nitrogen.

20. *Elements of Chemistry* (three volumes), translated by W. Nicholson, London, 1791.

CICERO, M. T., 106 B.C.–43 B.C.

21. *De Natura Deorum, II*, 23–28, translated by H. Rackham, Loeb Classical Library, 1948.

COLEMAN, Edward, 1765–1839. A surgeon, he was appointed at the age of twenty-eight Professor at the Veterinary College, London, despite his ignorance of the subject. A pupil of John Hunter, he wrote an essay on resuscitation, awarded the Humane Society's prize, based on his experiments on cats and dogs. His influence at the college was 'the greatest calamity the profession has ever experienced' (*Early History of Veterinary Literature*, vol. iii, Nineteenth century, 1800–23. London: 1930, pp. 13–25).

22. *A Dissertation on Natural and Suspended Respiration*, 2nd edition, London, 1802.

CRAWFORD, Adair, 1748–95. Physician and chemist, he made experiments on animal calorimetry.

23. Experiments and Observations on Animal Heat, 1st edition, London, 1779. Second edition, London, 1788. (All quotations are taken from the second edition.)

DAVY, Sir Humphry, 1778–1829.

24. *Researches, Chemical and Philosophical Chiefly concerning Nitrous Oxide, London, 1800* (Garnett, Thomas).
'*Outlines* of a Course of Lectures on Chemistry Delivered at the Royal Institution of Gt. Britain.' 1801. No author or editor listed.

DAVY, John, 1790–1868. Physiologist and anatomist, and younger brother of Sir Humphry Davy. He entered the army as surgeon and rose to the rank of Inspector-General of army hospitals. An F.R.S. (1834), he published over 150 memoirs and papers (*D.N.B.*).

25. 'An account of some Experiments on Animal Heat', *Philosophical Transactions of the Royal Society*, London, 1814, p. 390.

DELAROCHE, F.

26. 'On the Cause of Refrigeration observed in Animals exposed to a high Degree of Heat.' *Nicholson's Journal, 31*, 1812, p. 36.

DELAROCHE, F., and Berard, J.

27. 'Mémoire sur la Détermination de la Chaleur specifique des differents gaz.' *Annales de Chimie, LXXXV*, 1813, pp. 72–100, 113–82.

DESCARTES, R., 1596–1650.

28. 'De Homine Figuris et Latinitate Donatus a Florentia Schuyl', Leyden, 1664, translated by J. P. Mahaffy in *Descartes*, Philadelphia, 1881, etc.

DOUGLAS, Robert (dates unknown). A London physician, of whom little is known.

29. *An Essay Concerning the Generation of Heat in Animals*. London, 1747.

DRIESCH, H., 1867–1941.

29a. *The History and Theory of Vitalism*, translated by C. K. Ogden, London, 1914.

DUTROCHET, René Joachim Henri, 1776–1847. After a distinguished career as army-doctor during the Peninsular War, he devoted himself to physiology.

30. *Nouvelles recherches sur l'Endosmose et l'Exosmose, suivies de l'Application expérimentalé de ces actions physiques à la Solution du Problème de l'Irritabilité végétale*. Paris, 1828.

EDWARDS, William Frederic, 1777–1849. Born in Jamaica, he moved to Paris during the Revolution. He gained his doctorate in 1815. His experimental work on the effect of light on the body is of great importance.

31. *De l'Influence des Agents Physiques sur la Vie*, Paris, 1824, translated by Hodgkin and Fisher, 1832.

ELLIS, Daniel (dates unknown). Fellow of the Royal Society.

32. *An Inquiry into the Changes induced on atmospheric Air by the Germination of Seeds, the Vegetation of Plants, and the Respiration of Animals*, First edition, Edinburgh, 1807. Second edition, Edinburgh, 1811.

33. 'Reply to Dr. Bostock's Remarks on Mr. Ellis's Treatise on Respiration', *Edinburgh Medical Journal*, *4*, 1808, p. 325.

34. *Further Inquiries into the Changes induced on Atmospheric Air, etc.*, Edinburgh, 1811.

FAURE, J. L.,

35. *Claude Bernard*, Paris, 1925.

FAUST, E. D., p. 79.

36. 'Experiments and Observations of the Endesmose and Exosmose of Gases, and the Relation of these Phenomena with the Function of Respiration', *American Journal of Medical Science*, *5*, 1830.

FLOURENS, M., 1794–1867.

37. 'Memoir of Magendie', translated for *Smithsonian Report*, 1866, p. 91.

FRANKLIN, B., 1706–90.

38. 'Letter to J. Lining, 14th April, 1757', vol. 2, in *Works*, London, 1806.

GAY-LUSSAC, M., 1778–1850.

39. 'Observations Critiques sur les Phénomènes Chimiques de la Respiration', *Expérience*, Paris, 1844, *13*, p. 343.

HALLER, A. A., 1708–77.

40. *Primae Linae Physiologicae*, Goettingen, 1747, tranlated by Dr. Cullen, with notes by Wriseberg, Edinburgh, 1786.

HARVEY, W., 1578–1657.

41. *De Motu Cordis*, 1628. Translated by K. J. Franklin, Oxford, 1957.

42. *De Circulatione Sanguinis*, 1649. Translated by K. J. Franklin, Oxford, 1958.

VAN HELMONT, J. B., 1577 or 80–1644.

43. 'The Blas of Man', in *Oriatrike or Physick Refined*, London, 1662.

HENRY, J., 1799–1878.

44. Appendix to 'On Vitality', by H. H. Higgins, *Smithsonian Report*, 1866, p. 387.

HOOKE, R., 1635–1703.

45. *Posthumous Works*, London, 1705.

HUNTER, J., 1728–93.

46. 'Lectures on the Principles of Surgery', delivered in 1786–87. *Vol. I, Works.*

47. 'A Treatise on the Blood', 1793. *Vol. III, Works.* Both of these are in *Collected Works of John Hunter*, edited by J. Palmer, London, 1837.

47a. 'The Animal Oeconomy', *Vol. IV, Works.*

KUHN, T.S.

48. 'Energy Conservation as an Example of a Simultaneous Discovery' in *Critical Problems in the History of Science*, edited by M. Clagett, Madison, 1959.

JOULE, J. P., 1818–89.

49. 'On the Calorific Effects of Magneto-Electricity, and on the Mechanical Value of Heat', *Philosophical Magazine*, *XXIII*, 1843, p. 442.

LAVOISIER, A. L., 1743–94.

50. 'Mémoire sur la Nature du Principe qui se combine avec les Métaux pendant leur Calcination', *Memoire de l'Académie des Sciences* 1775.

51. 'Expériences sur la Respiration des Animaux, et sur les Changements qui arrivent à l'Air en passant par leur poumon', *Mémoire de L'Académie des Sciences*, 1777, published 1780. Reprinted in

the collection: *Les Maîtres de la Pensée scientifique, Paris,* 1920. Translated by T. Henry, see: *A Source Book in Animal Biology*, edited by T. S. Hall, New York, 1951.

52. 'Mémoire sur la Combustion en général', *Mémoires de l'Académie des Sciences,* 1777. Translated by Klickstein. See: *Source Book in Chemistry,* New York, 1952, p. 168.

33. *Traité Élémentaire de Chimie,* Parts I and II. Translated by R. Kerr, Edinburgh, 1790.

LAVOISIER, A. L., and Laplace, P. S. de.

54. 'Mémoire sur la Chaleur', *Mémoires de l'Académie des Sciences,* 1780, p. 355. Reprinted in *Les Maîtres de la Pensée Scientifique,* Paris, 1820. Translated by M. L. Gabriel in *Great Experiments in Biology,* edited by Gabriel and Fogel, 1955, pp. 85–93.

LAVOISIER, A. L., and Sequin.

55. 'Premier Mémoire sur la Respiration des Animaux', *Mémoires de l'Académie des Sciences,* 1789, p. 185.

56. 'Premier Mémoire sur la Transpiration des Animaux', *Mémoires de l'Académie des Sciences* , 1790, p. 77. Both 55 and 56 reprinted in the collection: *Les Maîtres de la Pensée Scientifique,* Paris, 1920.

LAWRENCE, W., 1783–67. Studied with Abernethy; surgeon at Bartholemew's Hospital, London; professor of surgery at the Royal College of Surgeons.

57. 'On Life.' *An Introduction to Comparative Anatomy and Physiology:* being two introductory Lectures delivered to the Royal College of Surgeons, London, 1816.

LESLIE, Peter Dugud, died 1782. Physician, Durham.

58. *A Philosophical Enquiry into the Cause of Animal Heat,* London, 1778.

LIEBIG, J., 1803–73.

59. *Animal Chemistry,* edited from the author's manuscript by W. Gregory. London, 1842. (Page references in this book are to be found in the Cambridge edition, 1943.)

McKIE, D.

60. 'Wöhler's Synthetic Urea and the Rejection of Vitalism: A Chemical Legend, *Nature, 153.* 1944, pp. 609.

61. *Antoine Lavoisier*, Second Edition. London, 1952. (also Schuman . . . New York, 1952).

McKie, D. and Heathcote, H. de V.

62. *The Discovery of Specific and Latent Heats*, London, 1935.

Magendie, F., 1783–1855.

63. 'Quelques Idées Générales sur les Phénomènes particulier aux Corps vivants', *Bulletin des Sciences Médicales*, 1809. p. 145.

64. *Précis Élémentaire de Physiologie*, 2nd edition, Paris, 1817.

64a. *Leçons sur les Fonctions et Maladies du Système Nerveux*, Paris, 1839–41.

64b. *Phénomènes physiques de la Vie*, Paris, 1842.

Magnus, M. G.

65. 'De la Présence de L'Oxigène, de l'Azote et de l'Acide carbonique dans le Sang, et sur la Théorie de la Respiration', *Annales des Sciences naturelles zoologiques*, 2nd series, *8*. 1837, p. 79.

Mayer, J. R., 1814–87.

66. 'Bemerkungen über die Kräfte der unbelebten Natur', *Annalen der Chemie und Pharmacie*, *42*, 1842.

67. 'The Mechanical Equivalent of Heat', in *The Correlation and Conservation of Forces*, edited by E. L. Youmans, New York, 1865.

Mayow, J., 1645–79.

68. *Tractatus Quinque*, Oxoni, 1674. Translated, The Alambic Club, Edinburgh, 1907.

Mitchell, J. K., 1793–1858.

69. 'On the Penetrativeness of Fluids', *American Journal of Medical Sciences*, 1830, p. 36.

Olmsted, J. M. D.

70. *Claude Bernard, Physiologist*, London, 1938.

71. *François Magendie*, New York, 1944.

72. *Brown-Sequard*, Baltimore, 1946.

Olmsted, J. M. D. and Olmsted, E. H.

73. *Claude Bernard and the Experimental Method in Medicine*, 2nd Edition, New York, 1952.

PALMER, Sir James Frederick, 1806–93. Practised for many years in London. Emigrated to Australia where he became president of the second Legislative Council of Victoria, and was knighted in 1857. His edition of the works of John Hunter appeared shortly before he left London (*Medical Times and Gazette*, 1871, *2*).

74. *The Works of John Hunter*, edited with notes by J. Palmer, London, 1837.

PEART, Edward, 1756?–1824. Physician, M.D., chiefly known for his works on physical and chemical theory. An independent observer of nature, whose ideas had less influence on his contemporaries than they deserved. He was an acute critic of Priestley and Lavoisier (*D.N.B.*).

75. *On Animal Heat*, London, 1788.

PHILIP, Alexander Philip Wilson, 1770–1851? Born and educated in Scotland, his original surname being Wilson. He was an F.R.S. and a Fellow of the Royal College of Physicians. Delivered the Culstonian Lectures in 1835. He was a popular physician and indefatigable worker, being one of the first to use the microscope in the study of inflammation. Little is known of his last years, Munk stating that 'his investments were injudicious'. He is thought to have died in Boulogne (*D.N.B.*).

76. *An Experimental Inquiry into the Laws of Vital Functions*, London, 1826.

POPPER, K. R.

76a. *The Logic of Scientific Discovery*, London, 1959.

PRIESTLEY, J., 1733–1804.

77. *Experiments and Observations on Different Kinds of Air*, Birmingham, 1774–77.

PRICHARD, James Cowles, 1786–1848. Physician, philologist, ethnologist, he practised in Bristol. His *Treatise on Insanity* (1835) was for long a standard work, while his *Researches into the physical history of man* contained in the second edition (1826) 'a remarkable anticipation of modern views on evolution' (Garrison-Morton). He was an F.R.S. and President of the Ethnological Society.

78. *A Review of the Doctrine of a Vital Principle*, London, 1829.

READ, J.

79. *Through Alchemy to Chemistry*, London, 1957.

RIGBY, Edward, *the elder*, 1747–1821. Began his studies under Priestley, and later studied in London. He earned a European reputation by his work on uterine haemorrhage. He was a practiser of vaccination in Norwich, where he had settled (*D.N.B.*).

80. *An Essay on the Theory of the Production of Animal Heat*, London, 1785.

SERVETUS, M., 1509–53.

81. *Christianismi Restitutio, 1553*. See also (for the passage quoted) *Servetus and Calvin* by R. Willis, London, 1877, p. 206.

SINGER, C.

82. *A Short History of Anatomy from the Greeks to Harvey*, Dover, 1957.

STEVENS, William, 1786–1868. Born and educated in Scotland, and a doctor of medicine of Copenhagen University (1820). Became noted for his treatment of cholera. (Hirsch and Hübotter, who give his dates wrongly as 1789–1862.)

83. *Observations on the Blood*, London, 1832.

VERNON, C.

84. 'A Sceptical Chemist', *The Listener*, April 3, 1958, p. 579.

WIGHTMAN, W. P. D.

85. *The Emergence of General Physiology*, Lecture delivered in the Queens University, Belfast, 1956.

WRISEBERG, Heinrich August, 1739–1808. Professor of anatomy, Göttingen, he is best remembered as the discoverer of the 'nervus intermedius' ('nerve of Wriseberg').

86. Notes to Dr. Cullen's translation of Haller's *Primae Linae Physiologicae*, see bibliographical reference 40.

HISTORY, PHILOSOPHY AND SOCIOLOGY OF SCIENCE

Classics, Staples and Precursors

An Arno Press Collection

Aliotta, [Antonio]. **The Idealistic Reaction Against Science.** 1914

Arago, [Dominique François Jean]. **Historical Eloge of James Watt.** 1839

Bavink, Bernhard. **The Natural Sciences.** 1932

Benjamin, Park. **A History of Electricity.** 1898

Bennett, Jesse Lee. **The Diffusion of Science.** 1942

[Bronfenbrenner], Ornstein, Martha. **The Role of Scientific Societies in the Seventeenth Century.** 1928

Bush, Vannevar. **Endless Horizons.** 1946

Campanella, Thomas. **The Defense of Galileo.** 1937

Carmichael, R. D. **The Logic of Discovery.** 1930

Caullery, Maurice. **French Science and its Principal Discoveries Since the Seventeenth Century.** [1934]

Caullery, Maurice. **Universities and Scientific Life in the United States.** 1922

Debates on the Decline of Science. 1975

de Beer, G. R. **Sir Hans Sloane and the British Museum.** 1953

Dissertations on the Progress of Knowledge. [1824]. 2 vols. in one

Euler, [Leonard]. **Letters of Euler.** 1833. 2 vols. in one

Flint, Robert. **Philosophy as Scientia Scientiarum and a History of Classifications of the Sciences.** 1904

Forke, Alfred. **The World-Conception of the Chinese.** 1925

Frank, Philipp. **Modern Science and its Philosophy.** 1949

The Freedom of Science. 1975

George, William H. **The Scientist in Action.** 1936

Goodfield, G. J. **The Growth of Scientific Physiology.** 1960

Graves, Robert Perceval. **Life of Sir William Rowan Hamilton.** 3 vols. 1882

Haldane, J. B. S. **Science and Everyday Life.** 1940

Hall, Daniel, et al. **The Frustration of Science.** 1935

Halley, Edmond. **Correspondence and Papers of Edmond Halley.** 1932

Jones, Bence. **The Royal Institution.** 1871

Kaplan, Norman. **Science and Society.** 1965

Levy, H. **The Universe of Science.** 1933

Marchant, James. **Alfred Russel Wallace.** 1916

McKie, Douglas and Niels H. de V. Heathcote. **The Discovery of Specific and Latent Heats.** 1935

Montagu, M. F. Ashley. **Studies and Essays in the History of Science and Learning.** [1944]

Morgan, John. **A Discourse Upon the Institution of Medical Schools in America.** 1765

Mottelay, Paul Fleury. **Bibliographical History of Electricity and Magnetism Chronologically Arranged.** 1922

Muir, M. M. Pattison. **A History of Chemical Theories and Laws.** 1907

National Council of American-Soviet Friendship. **Science in Soviet Russia: Papers Presented at Congress of American-Soviet Friendship.** 1944

Needham, Joseph. **A History of Embryology.** 1959

Needham, Joseph and Walter Pagel. **Background to Modern Science.** 1940

Osborn, Henry Fairfield. **From the Greeks to Darwin.** 1929

Partington, J[ames] R[iddick]. **Origins and Development of Applied Chemistry.** 1935

Polanyi, M[ichael]. **The Contempt of Freedom.** 1940

Priestley, Joseph. **Disquisitions Relating to Matter and Spirit.** 1777

Ray, John. **The Correspondence of John Ray.** 1848

Richet, Charles. **The Natural History of a Savant.** 1927

Schuster, Arthur. **The Progress of Physics During 33 Years (1875-1908).** 1911

Science, Internationalism and War. 1975

Selye, Hans. **From Dream to Discovery: On Being a Scientist.** 1964

Singer, Charles. **Studies in the History and Method of Science.** 1917/1921. 2 vols. in one

Smith, Edward. **The Life of Sir Joseph Banks.** 1911

Snow, A. J. **Matter and Gravity in Newton's Physical Philosophy.** 1926

Somerville, Mary. **On the Connexion of the Physical Sciences.** 1846

Thomson, J. J. **Recollections and Reflections.** 1936

Thomson, Thomas. **The History of Chemistry.** 1830/31

Underwood, E. Ashworth. **Science, Medicine and History.** 2 vols. 1953

Visher, Stephen Sargent. **Scientists Starred 1903-1943 in American Men of Science.** 1947

Von Humboldt, Alexander. **Views of Nature: Or Contemplations on the Sublime Phenomena of Creation.** 1850

Von Meyer, Ernst. **A History of Chemistry from Earliest Times to the Present Day.** 1891

Walker, Helen M. **Studies in the History of Statistical Method.** 1929

Watson, David Lindsay. **Scientists Are Human.** 1938

Weld, Charles Richard. **A History of the Royal Society.** 1848. 2 vols. in one

Wilson, George. **The Life of the Honorable Henry Cavendish.** 1851